Neurosurgical Procedures - Understanding and Advancements

Edited by Amit Agrawal

Published in London, United Kingdom

Neurosurgical Procedures - Understanding and Advancements
http://dx.doi.org/10.5772/intechopen.1010467
Edited by Amit Agrawal

Contributors
Abdulla Illeyan, Alicia Dupre, Ali Msheik, Amit Agrawal, Amr El Mohamad, Andrew Treister, Angela Leanca, Bea Drucks, Conor Gillespie, Eduardo Bertolini, Egor Markin, Helen Fowler, Hubertus Maximilian Mehdorn, Ilshat Gaysin, Isabel Luebbing, Kazim Mohammed, Manel Santafè Colomina, Muath Hussein, Muhammad Mohsin Khan, Nasser Alsaad, Ruben Peralta, Valeriu Timirgaz, Yevgeny Karepov, Youssef Chedid, Zeinab Al Mokdad

First published in London, United Kingdom, 2026 by IntechOpen
IntechOpen is the global imprint of INTECHOPEN LIMITED, registered in England and Wales, registration number: 11086078, 167-169 Great Portland Street, London, W1W 5PF, United Kingdom

For EU product safety concerns: IN TECH d.o.o., Prolaz Marije Krucifikse Kozulić 3, 51000 Rijeka, Croatia, info@intechopen.com or visit our website at intechopen.com.

British Library Cataloguing-in-Publication Data
A catalogue record for this book is available from the British Library

Neurosurgical Procedures - Understanding and Advancements
Edited by Amit Agrawal
p. cm.
Print ISBN 978-1-80631-853-7
Online ISBN 978-1-80631-852-0
eBook (PDF) ISBN 978-1-80631-854-4

Meet the editor

Dr. Agrawal completed his neurosurgery training at the National Institute of Mental Health and Neurosciences, Bangalore, India, in 2003. Dr. Agrawal is a self-motivated, enthusiastic, and results-oriented professional with over 21 years of extensive experience in research and development, as well as teaching and mentoring in neurosurgery. He is proficient in managing and leading teams to run successful processes and has experience developing procedures and service standards of excellence. He has attended and participated in many international and national-level symposiums and conferences and delivered lectures on various topics. He has published more than 1,000 articles in the medical field, covering various topics in national and international journals. His expertise lies in identifying training needs, designing training modules, and executing them while working with limited resources. He has excellent communication, presentation, and interpersonal skills with proven abilities in teaching and training for various academic and professional courses. Presently, he is working at the All-India Institute of Medical Sciences, Bhopal, Madhya Pradesh (India).

Contents

Preface

Neurosurgical Procedures – Understanding and Advancements brings together contributions that examine several of the most significant developments currently influencing neurosurgical practice, spanning neurocritical care, minimally invasive intervention, functional neurosurgery, and healthcare systems innovation. The volume is organized into three sections: precision care in neurosurgical disease, minimally invasive and functional interventions, and technology-driven approaches to modern neurosurgical practice.

The opening section, *Precision Care in Neurosurgical Disease: Monitoring, Emergencies, and Outcomes*, addresses advances in neurocritical care, prognostic assessment, and evidence-based management of complex neurological conditions. Chapters examine multimodal neuromonitoring in subarachnoid hemorrhage, neuro-oncologic emergencies, prognostic evaluation of pediatric brain tumors associated with obstructive hydrocephalus, and contemporary approaches to traumatic cervical spine injuries.

The discussion then shifts to *Minimally Invasive, Endovascular, and Functional Neurosurgery*, examining how less-invasive, precision-guided interventions are reshaping clinical practice. Topics include evidence-based comparisons of transcallosal and endoscopic approaches for third ventricular colloid cysts, as well as functional neurosurgical therapies, such as deep brain stimulation and MRI-guided focused ultrasound, for movement disorders.

The final section, *Technology, Systems, and the Future of Neurosurgical Practice,* broadens the focus beyond individual procedures to the systems that support neurosurgical care. Chapters address patient positioning and operative optimization, intelligent perioperative pathways, artificial intelligence–enabled clinical decision support, predictive risk stratification, health systems strengthening, and equitable access to neurosurgical services.

Although the topics covered in this volume span diverse areas of neurosurgery, a common theme emerges throughout: the pursuit of more precise, evidence-informed, and patient-centered care. The chapters reflect advances not only in operative technique but also in monitoring, prognostication, multidisciplinary decision-making, and healthcare delivery. Collectively, they illustrate how contemporary neurosurgery increasingly extends beyond the operating room, integrating technological innovation, critical care, functional preservation, and system-level thinking to improve outcomes for patients and populations alike.

Amit Agrawal
Department of Neurosurgery,
All India Institute of Medical Sciences,
Saket Naga, Bhopal, Madhya Pradesh, India

Section 1

Precision Care in Neurosurgical Disease: Monitoring, Emergencies, and Outcomes

Chapter 1

Introductory Chapter: Neurosurgical Paradigm – from Technical Mastery to Systemic Precision

Amit Agrawal

1. Introduction

The progress in neurosurgery has changed from concepts that were historically considered unrealistic to the current stage where these are incorporated into standard clinical practice. [1] The open spinal surgery approach, which was considered the conventional and standard treatment for spinal disorders, [2] is now gradually being replaced by minimally invasive spine surgery (MISS) methods for various spinal diseases. [3–5] The concept of MISS encompasses minimizing disruption to surrounding tissues while allowing adequate exposure and treatment of spinal pathologies and reducing the morbidity associated with open spine surgeries. [2] Other advancements that have expanded the precision and safety margins of neurosurgical procedures include stereotactic guidance, endoscopic visualization, navigation systems, intraoperative imaging, [6] and augmented reality (AR)-based technologies to enhance the surgical viewing experience. [7] More recently, the exoscope has provided an alternative to conventional surgical microscopes, offering a visualization platform consisting of a small stereo camera and a 4 K three-dimensional display to reconstruct a three-dimensional stereoscopic image. [8]

2. Nanoneurosurgery and targeted therapeutics

Nanoneurosurgery presents the possibility of developing treatments that extend beyond disease control to include tissue regeneration, restoration of function, and the development of new surgical capabilities. [1, 9–11] Targeted nanoparticle-based therapeutic strategies are capable of delivering molecules across the blood-brain barrier. [10, 12, 13]

3. Advances in neuroimaging and tractography

Over a period of time, neuroimaging techniques have evolved significantly. Diffusion-weighted imaging (DWI)-based tractography helps in identifying the structural organization of white matter (WM), and in combination with bundle

 IntechOpen

segmentation, it is possible to study functionally critical connections. [14] The identification of white matter connections helps preserve functionally essential WM bundles during tumor resection and consequently preserve cognitive function. [15] However, the use of tractographic data requires knowledge of fiber characteristics, including the effects of brain shift, crossing fibers, and other anatomically relevant metrics. [14] Virtual training modules facilitate microdissection training and can help provide direct visualization of WM anatomy and dynamic interaction with tractography datasets, facilitating the training of clinicians and neurosurgeons and improving their proficiency in WM anatomy. [14]

4. High-grade glioma surgery and functional preservation

High-grade gliomas (HGGs; World Health Organization [WHO] III–IV) comprise approximately 50% of glioma cases. [16] Advancements in multimodal treatment approaches, which include molecular characterization and aggressive surgical resection followed by advanced therapeutic strategies, together with evidence supporting greater extent of resection, have contributed to improved outcomes in glioma management. [17–21] However, the mean survival remains 12–14 months, with fewer than 10% of patients surviving 5 years after diagnosis. [22, 23] Modern methods such as the evaluation of white matter (WM) pathways and their spatial relationship with the tumor [14] and awake mapping techniques [24] have further improved surgical planning and tumor resection.

5. Evolution of neurovascular and aneurysm management

Neurovascular practice has evolved into specialized and multidisciplinary treatment approaches. [25, 26] Surgical clipping of a ruptured aneurysm is considered a highly stress-inducing procedure and is reserved for selected anatomical and clinical situations. [25–27] Over time, the management of complex aneurysms has evolved into a multidisciplinary team approach where combined bypass and endovascular approaches have been developed as less invasive and effective treatment options. [25, 26] This has been made possible by the availability of skilled and motivated neuroradiologists, a multidisciplinary team approach, and close collaboration with neurosurgeons. [27]

6. Connectome-guided and network-centric neurosurgery

Connectome-guided neurosurgery, which represents a transition from lesion-centric to network-centric neurosurgery, characterizes brain tumors as network disorders and aims to preserve brain function during tumor resection. [24] The connectome, which represents the point-to-point mapping of neuronal circuits within the brain, [28] has further facilitated the transition of brain mapping from a localizationist model to a network-based approach. [29] These networks can be mapped and studied with preoperative functional MRI [30] and facilitate the mapping of relationships between glial tumors and structural as well as functional neural networks. [31] The understanding of this relationship helps in operative planning and the assessment of postoperative outcomes. [29]

7. Functional Mapping and Intraoperative Technologies

Awake surgery with direct electrical stimulation (DES) for mapping language and higher cognitive functions and defining the subcortical boundaries of resection, [24, 32, 33] intraoperative electrocorticography (ECoG) with gamma band modulation for achieving safe resection margins during awake craniotomy, [34] and the use of magnetoencephalography (MEG) as a direct measure of neuronal activity and synaptic currents [35] are among other innovative methods that are increasingly being used.

Overall, the continued advances in neurosurgery, including precision-guided interventions, technological innovation, multidisciplinary collaboration, and educational adaptability, are reflected in contemporary patient-care frameworks. [1, 29] The core objective remains collaboration with professionals across disciplines to enhance both surgical practice and overall team performance, facilitate knowledge sharing and decision-making, and improve the management of patients and underlying diseases, with the goal of achieving better overall outcomes. [27]

Acknowledgements

The author acknowledges the use of Grammarly for language polishing of the manuscript.

Author details

Amit Agrawal
Department of Neurosurgery, All India Institute of Medical Sciences Bhopal, Bhopal, India

*Address all correspondence to: dramitagrawal@gmail.com; dramitagrawal@hotmail.com

References

[1] Qureshi P. Nanoneurosurgery: Pioneering a paradigm shift for the 21st century. Surgical Neurology International. 2026;**17**:25. DOI: 10.25259/SNI_981_2025

[2] Smith ZA, Fessler RG. Paradigm changes in spine surgery: Evolution of minimally invasive techniques. Nature Reviews Neurology. 2012;**8**(8):443–450. DOI: 10.1038/nrneurol.2012.110

[3] Khoo LT, Fessler RG. Microendoscopic decompressive laminotomy for the treatment of lumbar stenosis. Neurosurgery. 2002;**51**(5 Suppl):S146–154

[4] Kim DH, O'Toole JE, Ogden AT, et al. Minimally invasive posterolateral thoracic corpectomy: Cadaveric feasibility study and report of four clinical cases. Neurosurgery. 2009;**64**(4):746–752. DOI: 10.1227/01.NEU.0000340783.83964.27

[5] Kan P, Schmidt MH. Minimally invasive thoracoscopic approach for anterior decompression and stabilization of metastatic spine disease. Neurosurgical Focus. 2008;**25**(2):E8. DOI: 10.3171/FOC/2008/25/8/E8

[6] Gadjradj PS, Rubinstein SM, Peul WC, et al. Full endoscopic versus open discectomy for sciatica: Randomised controlled non-inferiority trial. Bmj. 2022;**376**:e065846

[7] Hey G, Guyot M, Carter A, Lucke-Wold B. Augmented Reality in Neurosurgery: A New Paradigm for Training. Medicina (Kaunas). 2023;**59**(10):1721. DOI: 10.3390/medicina59101721

[8] Kinoshita M, Kijima N, Kagawa N, Kishima H. The Exoscope Paradigm Shift in Neurosurgery. Japanese Journal of Neurosurgery. 2021;**30**(3):199–207. DOI: 10.7887/jcns.30.199

[9] Klein AL, Nugent G, Cavendish J, et al. Nanoparticles as a Tool in Neuro-Oncology Theranostics. Pharmaceutics. 2021;**13**(7):948. DOI: 10.3390/pharmaceutics13070948

[10] Sagar G, Wilson B, Kannoth Mukundan G, Divekar K, Jenita JL. Nanomedicine for Treating Major Brain Diseases: Advances and Future Directions. Molecular Pharmaceutics. 2025;**22**(8):4413–4434. DOI: 10.1021/acs.molpharmaceut.5c00277

[11] Ye Q, Zhang J, Wang X, et al. Recent advances in nanomedicine strategies for nervous system injuries biomolecular regeneration. Materials & Design. 2026;**261**:115333

[12] Ferrari M. Cancer nanotechnology: Opportunities and challenges. Nature Reviews Cancer. 2005;**5**(3):161–171. DOI: 10.1038/nrc1566

[13] Kochman U, Sitka H, Kuzniar J, et al. Targeted Nanoparticles for Drug Delivery Across the Blood-Brain Barrier in Early and Late Stages of Alzheimer's Disease: A Review. Molecular Neurobiology. 2025;**63**(1):75. DOI: 10.1007/s12035-025-05417-z

[14] Sarubbo S, Vavassori L, Zigiotto L, et al. Changing the Paradigm for Tractography Segmentation in Neurosurgery: Validation of a Streamline-Based Approach. Brain Sciences. 2024;**14**(12):1232. DOI: 10.3390/brainsci14121232

[15] de Schotten M T, Forkel SJ. The emergent properties of the connected brain. Science. 2022;**378**(6619):505–510. DOI: 10.1126/science.abq2591

[16] Ohgaki H. Epidemiology of brain tumors. Methods in Molecular Biology. 2009;**472**:323–342. DOI: 10.1007/978-1-60327-492-0_14

[17] Jackson C, Choi J, Khalafallah AM, et al. A systematic review and meta-analysis of supratotal versus gross total resection for glioblastoma. Journal of Neuro-Oncology. 2020;**148** (3):419–431. DOI: 10.1007/s11060-020-03556-y

[18] Zigiotto L, Annicchiarico L, Corsini F, et al. Effects of supra-total resection in neurocognitive and oncological outcome of high-grade gliomas comparing asleep and awake surgery. Journal of Neuro-Oncology. 2020;**148**(1):97–108. DOI: 10.1007/s11060-020-03494-9

[19] Sanai N, Polley MY, McDermott MW, Parsa AT, Berger MS. An extent of resection threshold for newly diagnosed glioblastomas. Journal of Neurosurgery. 2011;**115**(1):3–8. DOI: 10.3171/2011.2.jns10998

[20] Capelle L, Fontaine D, Mandonnet E, et al. Spontaneous and therapeutic prognostic factors in adult hemispheric World Health Organization Grade II gliomas: A series of 1097 cases: Clinical article. Journal of Neurosurgery. 2013;**118**(6):1157–1168. DOI: 10.3171/2013.1.JNS121

[21] Aziz PA, Memon SF, Hussain M, et al. Supratotal Resection: An Emerging Concept of Glioblastoma Multiforme Surgery-Systematic Review And Meta-Analysis. World Neurosurgery. 2023;**179**:e46–e55

[22] Lacroix M, Abi-Said D, Fourney DR, et al. A multivariate analysis of 416 patients with glioblastoma multiforme: Prognosis, extent of resection, and survival. Journal of Neurosurgery. 2001;**95**(2):190–198. DOI: 10.3171/jns.2001.95.2.0190

[23] McGirt MJ, Than KD, Weingart JD, et al. Gliadel (BCNU) wafer plus concomitant temozolomide therapy after primary resection of glioblastoma multiforme. Journal of Neurosurgery. 2009;**110**(3):583–588. DOI: 10.3171/2008.5.17557

[24] Patel T, Yousuf Ibrahim H, Abbas M, et al. Emerging paradigms in connectome-guided neurosurgery for central nervous system tumors: A narrative review of structural and functional connectomics. Annals of Medicine & Surgery. 2026;**00**:1–8

[25] Tayebi Meybodi A, Huang W, Benet A, Kola O, Lawton MT. Bypass surgery for complex middle cerebral artery aneurysms: An algorithmic approach to revascularization. Journal of Neurosurgery. 2017;**127**(3):463–479. DOI: 10.3171/2016.7.JNS16772

[26] Murayama Y, Arakawa H, Ishibashi T, et al. Combined surgical and endovascular treatment of complex cerebrovascular diseases in the hybrid operating room. Journal of NeuroInterventional Surgery. 2013;**5** (5):489–493. DOI: 10.1136/neurintsurg-2012-010382

[27] Somma T, Ius T, Certo F, et al. From the Champion to the Team: New Treatment Paradigms in Contemporary Neurosurgery. World Neurosurgery. 2019;**131**:141–148

[28] Toga AW, Clark KA, Thompson PM, Shattuck DW, Van

Horn JD. Mapping the human connectome. Neurosurgery. 2012;**71**(1):1–5. DOI: 10.1227/NEU.0b013e318258e9ff

[29] Catalino MP, Yao S, Green DL, Laws ER, Golby AJ, Tie Y. Mapping cognitive and emotional networks in neurosurgical patients using resting-state functional magnetic resonance imaging. Neurosurgical Focus. 2020;**48**(2):E9. DOI: 10.3171/2019.11.FOCUS19773

[30] Craddock RC, Tungaraza RL, Milham MP. Connectomics and new approaches for analyzing human brain functional connectivity. Gigascience. 2015;**4**:13. DOI: 10.1186/s13742-015-0045-x

[31] Samuel N, Vetkas A, Pancholi A, et al. A Network-Based Approach to Glioma Surgery: Insights from Functional Neurosurgery. Cancers (Basel). 2021;**13**(23):6127. DOI: 10.3390/cancers13236127

[32] Sanai N, Berger MS. Intraoperative stimulation techniques for functional pathway preservation and glioma resection. Neurosurgical Focus. 2010;**28**(2):E1. DOI: 10.3171/2009.12.FOCUS09266

[33] Sanai N, Mirzadeh Z, Berger MS. Functional outcome after language mapping for glioma resection. New England Journal of Medicine. 2008;**358**(1):18–27. DOI: 10.1056/NEJMoa067819

[34] Noll KR, Asman P, Tasnim I, et al. Intraoperative language mapping guided by real-time visualization of gamma band modulation electrocorticograms: Case report and proof of concept. Neuro-oncology Practice. 2024;**11**(1):92–100. DOI: 10.1093/nop/npad059

[35] Fred AL, Kumar SN, Kumar Haridhas A, et al. A Brief Introduction to Magnetoencephalography (MEG) and Its Clinical Applications. Brain Sciences. 2022;**12**(6):788. DOI: 10.3390/brainsci12060788

Chapter 2

Beyond the Aneurysm: Advanced Neuromonitoring Strategies in Subarachnoid Hemorrhage Care

Manel Santafè Colomina

Abstract

Aneurysmal subarachnoid hemorrhage (aSAH) remains a devastating neurological condition where clinical outcomes are dictated not only by the primary bleed but also by a complex cascade of secondary injuries, including early brain injury (EBI) and delayed cerebral ischemia (DCI). Traditional reactive management based on generic thresholds often fails to detect incipient cerebral distress, particularly in comatose or sedated patients where the clinical examination is non-informative. Consequently, modern neurocritical care has transitioned toward a proactive multimodal monitoring (MMM) paradigm, integrating diverse physiological signals to deliver personalized care. This chapter reviews essential MMM modalities through an integrative framework. We first examine foundational metrics like intracranial pressure (ICP)—emphasizing waveform morphology and intracranial compliance (ICC) over absolute thresholds—and cerebral perfusion pressure (CPP). Hemodynamic context is provided via Transcranial Doppler (TCD), a critical non-invasive tool for screening vasospasm. Oxygenation monitoring is discussed by comparing the invasive gold standard, brain tissue oxygen tension (PbtO2), with non-invasive Near-Infrared Spectroscopy (NIRS), highlighting their lack of interchangeability. To assess cellular bioenergetics, we review cerebral microdialysis (CMD), which identifies metabolic energy failure and can predict DCI up to 18 hours before clinical manifestation. Finally, the role of continuous EEG is explored for detecting non-convulsive electrical crises mechanistically linked to metabolic distress. By adopting a tiered, stepwise approach to monitoring, clinicians can move beyond "one-size-fits-all" targets toward a physiology-guided strategy that anticipates secondary insults, ultimately facilitating individualized interventions in the most vulnerable aSAH populations.

Keywords: aneurysmal subarachnoid hemorrhage, multimodal neuromonitoring, delayed cerebral ischemia, cerebral microdialysis, brain tissue oxygenation

1. Introduction: Overview of aneurysmal subarachnoid hemorrhage

Aneurysmal subarachnoid hemorrhage (aSAH) remains one of the most devastating forms of stroke, accounting for approximately 5% of all stroke cases. It represents a major global public health problem, with an estimated worldwide incidence of approximately 6.1 cases per 100,000 persons per year (around 27,000 patients annually) and a prevalence of about 8.09 million cases [1]. Mortality rates remain high, with approximately 50% of patients dying within 30 days and nearly half of the survivors experiencing permanent disability [1, 2].

The clinical course of aSAH can be divided into an early brain injury (EBI) phase, occurring within the first 72 hours after ictus, and a late brain injury phase characterized by the development of delayed cerebral ischemia (DCI), which may occur between days 4 and 12.

The EBI is characterized by the extravasation of blood into the subarachnoid space, causing a sudden increase in intracranial pressure (intracranial hypertension (ICH)), leading to an abrupt reduction in cerebral blood flow (CBF) and pressure-related brain injury. When an intraparenchymal component is present, the hemorrhage may also cause direct neuronal destruction. During this early phase, cerebral complications include secondary cerebral edema, epileptogenic processes due to blood irritation and parenchymal injury, obstructive hydrocephalus, and rebleeding, which is associated with increased mortality [1, 3]. Systemic complications result from a catecholaminergic surge that may trigger a systemic stress response [4, 5] and multiorgan dysfunction. Stress-induced cardiomyopathy is a prominent manifestation [6], ranging from electrocardiographic changes and transient elevation of cardiac biomarkers to regional wall motion abnormalities or cardiogenic shock [7, 8], and neurogenic pulmonary edema [9].

During this phase, close systemic and cerebral monitoring is required. Intracranial pressure may be elevated due to both the initial injury and secondary cerebral edema, leading to impaired CBF and ultimately compromised cerebral perfusion.

Between days 4 and 12, late brain injury could appear. DCI is defined as the development of an established cerebral infarction detected by CT or MRI, or confirmed at clinical autopsy, which cannot be attributed to the initial aneurysm occlusion in comatose patients. In patients with preserved or high levels of consciousness, DCI is defined by the appearance of new neurological focal deficits – such as hemiparesis, aphasia, apraxia, hemianopia, or neglect – or by a sustained decrease in the level of consciousness (a reduction of at least two points on the Glasgow Coma Scale lasting for at least one hour), not attributable to other medical causes such as epileptic seizures, fever, hydrocephalus, electrolyte disturbances, or other systemic conditions. DCI affects approximately one-third of patients and is a primary contributor to poor long-term outcomes [10].

DCI is a multifactorial process [11–13] involving vascular dysfunction, microthrombosis, neuroinflammation, and cortical spreading depolarizations. Large-vessel vasospasm, together with impaired autoregulation and endothelial dysfunction, alters macro and microcirculation but is neither necessary nor sufficient to cause DCI [14]. Clinically relevant hypoperfusion may arise from microvascular failure and angiographically occult microthrombi [15]. Hemorrhage-induced neuroinflammation, driven by neuronal and endothelial injury and blood degradation products [16], promotes oxidative stress, blood–brain barrier disruption, edema, vasoconstriction, and programmed cell death [17], thereby facilitating sustained hypoperfusion and cortical vulnerability. In this setting, cortical spreading depolarizations increase metabolic demand in hypoperfused tissue, exacerbating ischemic injury [18, 19].

The assessment of neurological status through clinical examination remains a fundamental pillar in aSAH management, yet its application presents critical limitations as a solitary surveillance tool. In the context of neurocritical care, reliance on physical examination alone can delay the detection of reversible secondary complications, thereby compromising the final patient prognosis.

In patients presenting with poor-grade aSAH (World Federation of Neurosurgical Societies [WFNS] IV–V), the clinical examination is inherently limited by the patient's initial deep coma or lack of responsiveness [20]. This limitation is further aggravated by the frequent need for mechanical ventilation and deep sedation to control intracranial pressure (ICP) or to protect the airway, which masks informative motor and cognitive responses. The use of anesthetic agents essentially eliminates the possibility of performing a detailed clinical assessment [21]. Furthermore, the practice of interrupting sedation for "neurological windows" is often contraindicated in patients with raised ICP, as the resulting agitation or loss of ICP control can be deleterious [22].

To bridge these gaps, multimodal neuromonitoring provides a continuous, real-time assessment of cerebral hemodynamics, oxygenation, and metabolism [22]. By integrating data from ICP, transcranial Doppler (TCD), brain tissue oxygenation, and cerebral microdialysis, clinicians can identify incipient cerebral distress – a state of "energy failure" at the cellular level – before permanent infarction occurs [23].

For multimodal neuromonitoring assessment, tools can be classified into the following categories:

- **Hemodynamic monitoring tools:** TCD and ICP monitoring.
- **Oxygenation and metabolic monitoring tools:** brain tissue oxygen tension ($PbtO_2$), near-infrared spectroscopy (NIRS), and cerebral microdialysis (CMD).
- **Electrical activity monitoring tools:** electroencephalography (EEG)

1.1 Methods

This chapter is based on a structured narrative review of the literature focusing on advanced neuromonitoring strategies in aneurysmal subarachnoid hemorrhage (aSAH). A comprehensive search of the PubMed/MEDLINE database was performed to identify relevant publications, using combinations of the following keywords: "subarachnoid hemorrhage," "multimodal neuromonitoring," "intracranial pressure," "cerebral perfusion pressure," "brain tissue oxygenation," "PbtO2," "cerebral microdialysis," "transcranial Doppler," and "electroencephalography." Priority was given to:

- International guidelines and consensus statements from recognized scientific societies.
- Randomized controlled trials and prospective cohort studies.
- High-quality observational studies.

- Expert consensus documents are published in high-impact, peer-reviewed journals.

- This review was not conducted as a formal systematic review with meta-analysis; therefore, no quantitative synthesis of data was performed.

- The authors acknowledge the use of ChatGPT (OpenAI, San Francisco, CA, USA) and Grammarly (Grammarly Inc., San Francisco, CA, USA) for language polishing and stylistic editing of the manuscript. All scientific content, interpretation, and final revisions were performed and validated by the authors. The authors take full responsibility for the content of this chapter.

2. Multimodal neuromonitoring

2.1 Hemodynamic flow monitoring

2.1.1 Transcranial Doppler

TCD ultrasonography is a noninvasive, bedside technology that utilizes ultrasound waves to monitor blood flow velocities in the large basal intracranial arteries [22]. The technique is based on the Doppler shift, where ultrasound waves are reflected off moving red blood cells, with the resulting frequency change being proportional to the velocity of those cells. While TCD measures relative rather than absolute CBF, a linear relationship exists between flow velocity and CBF as long as the cross-sectional area of the monitored vessel and the angle of insonation remain constant [24].

The primary metrics obtained using TCD are mean flow velocity (MFV) and the pulsatility index (PI). In the context of aSAH, MFV thresholds in the middle cerebral artery [25] are commonly used to screen for vasospasm; an MCA MFV >120 cm/s may indicate the presence of large-vessel vasospasm at this level.

To distinguish between true vasospasm (vessel narrowing) and hyperemia (increased global blood flow), the Lindegaard ratio is useful [26]. This is calculated by dividing the MFV of the MCA by the MFV of the extracranial internal carotid artery. A ratio >3 indicates moderate vasospasm, while a ratio >6 is suggestive of severe vasospasm [24].

Furthermore, the pulsatility index (PI) – a calculation derived from peak systolic, end-diastolic, and mean flow velocities – serves as a surrogate for distal cerebrovascular resistance. A PI >1.2 is frequently used as a noninvasive indicator of ICH [27]. However, it is critical to acknowledge that the absolute accuracy of TCD-based ICP estimation is approximately ± 15 mmHg, which limits its role to screening and longitudinal trend monitoring rather than serving as a definitive substitute for invasive gold-standard measurements [22].

TCD can be used to estimate ICP (ICPtcd) noninvasively using CBF velocity waveforms and validated threshold models. Cerebral perfusion pressure (CPP) can be estimated as CPPe = MAP × (Vd/Vm) + 14, where MAP is mean arterial pressure, Vd is diastolic flow velocity, and Vm is MFV in the middle cerebral artery. Estimated ICP is then calculated as ICPtcd = MAP – CPPe. ICPtcd has demonstrated a high negative predictive value for ruling out ICH compared with invasive ICP monitoring, suggesting its utility as a screening tool when invasive methods are unavailable or contraindicated. Although it does not replace direct measurement, TCD-based ICP estimation may support clinical decision-making by identifying patients at low risk of elevated ICP in acute brain injury [28].

TCD is highly sensitive for detecting vasospasm, but beyond absolute thresholds, trend monitoring is vital for clinical decision-making. Daily examinations that reveal an increase in MCA velocities of >50 cm/s over a 24-hour period are considered highly predictive of vasospasm [10, 29]. Early abnormalities in these indices have been shown to predict neurological deterioration with high accuracy, providing a window for "intention-to-treat" interventions such as induced hypertension or endovascular angioplasty [30].

The primary strengths of TCD are its noninvasiveness, portability, and capability for real-time assessment of cerebral hemodynamics at the patient's bedside. It is uniquely suited for serial monitoring during therapeutic interventions [27].

However, this technique has several important limitations, including high operator dependency, limited feasibility due to inadequate acoustic windows in approximately 10% of patients, reduced sensitivity outside the middle cerebral artery territory, and significant interobserver variability, which may hinder reliable interpretation of longitudinal measurements [22, 27].

2.1.2 Hemodynamic monitoring tool: ICP and cerebral perfusion pressure monitoring

The early phase of management of aSAH is fundamentally centered on the control of ICH and the optimization of cerebral energetics. Following the rupture of a cerebral aneurysm, blood accumulates within the subarachnoid space, triggering an immediate rise in ICP [31]. This acute event initiates the EBI phase, characterized by a sudden reduction in the pressure gradient required for CBF. According to the Monro-Kellie doctrine, the skull is a rigid compartment containing brain parenchyma, blood, and cerebrospinal fluid (CSF); any increase in volume from a space-occupying lesion, such as an intraparenchymal hematoma or cerebral edema, must be compensated by the displacement of CSF or venous blood [22, 27]. When these compensatory reserves – defined as intracranial compliance – are exhausted, ICP increases exponentially, leading to structural displacement, brain tissue herniation, and global ischemia.

CPP is the physiological surrogate for the pressure gradient driving CBF and is calculated as the difference between mean arterial pressure (MAP) and (ICP (CPP = MAP – ICP) [32]. Maintaining an adequate CPP is critical for preventing secondary brain injury and mitigating the effects of DCI. In the delayed phase of aSAH, cerebrovascular resistance rises due to vasospasm and microvascular dysfunction, while autoregulatory capacity is frequently impaired.

Clinicians often rely on fixed thresholds, largely extrapolated from traumatic brain injury (TBI) protocols, which generally suggest maintaining ICP below 20–22 mmHg and CPP between 60 and 70 mmHg [33]. However, aSAH is a distinct disease entity characterized by unique stressors such as hydrocephalus and vasospasm [32]. Rigid adherence to universal thresholds ignores the heterogeneity of brain injury and the temporal shifts in pathophysiology [22, 27]. For example, while a CPP of 60–70 mmHg is acceptable in many TBI cases, aSAH patients often require a CPP above 70 mmHg or even up to 90–100 mmHg during vasospasm phases to prevent clinical deterioration and improve metabolic profiles [32].

The concept of optimal CPP (CPPopt) seeks to identify the CPP range associated with preserved autoregulation in a given patient. It is determined by monitoring the pressure reactivity index (PRx), a moving correlation between MAP and ICP; a negative PRx indicates intact autoregulation, and the CPPopt is found at the point where PRx is most negative [22, 32].

ICP and CPP monitoring serve as the foundation of neurocritical care, but they are insufficient in isolation. Significant metabolic distress and electrical "crises," such as non-convulsive seizures (NCS) and cortical spreading depolarizations, can occur even when ICP and CPP remain within "normal" ranges [25, 27]. Advanced multimodal monitoring integrates ICP/CPP data with brain tissue oxygenation and CMD [10, 34, 35].

2.2 Brain oxygenation monitoring: Pbto2 and NIRS

The monitoring of cerebral oxygenation is a vital component of the multimodal paradigm in aSAH care, as cerebral hypoxia frequently occurs even in the absence of ICH [34]. While traditional management focused on pressure thresholds, modern neurocritical care emphasizes the balance between oxygen delivery and consumption to prevent secondary ischemic injury [27].

Cerebral tissue oxygenation reflects the balance between oxygen delivery and consumption. Supply is determined by CBF – which depends on CPP and vascular autoregulation – and arterial oxygen content, which is a product of hemoglobin levels and oxygen saturation [34]. Consumption increases significantly during agitation, fever, or seizures, potentially precipitating "hypermetabolic hypoxia" even if delivery is technically adequate [36].

The $PbtO_2$ sensor provides continuous, focal measurements of brain parenchymal oxygen tension, with normal values generally considered to range between 20–35 mmHg [37]. Low $PbtO_2$ values have been associated with increased morbidity and mortality, particularly when values fall below 10 mmHg. A threshold below 20 mmHg is commonly interpreted as indicative of impaired tissue oxygenation and should prompt consideration of therapeutic interventions aimed at optimizing oxygen delivery [38, 39].

Currently, two monitoring systems are available for measuring $PbtO_2$. Both systems are validated and provide reliable $PbtO_2$ measurements [40].

The Licox® system is based on a modified Clark-type electrode, consisting of a probe with an oxygen-permeable semipermeable membrane covering a cathode and an anode immersed in an electrolyte solution. Dissolved oxygen is reduced at the cathode, generating an electrical current proportional to the local oxygen partial pressure [41, 42].

More recently, the Raumedic® Neurovent-PTO system has been introduced, which is based on an optical luminescence principle. Light is emitted by a luminescent agent (ruthenium) located at the probe tip, and the reduction in emitted light intensity, caused by oxygen-dependent quenching, is proportional to the partial pressure of oxygen in the surrounding medium [43].

Because multiple physiological and pathological factors may influence $PbtO_2$ values – including $PaCO_2$ variations, fever, and epileptic seizures – $PbtO_2$ should not be interpreted in isolation [39].

When $PbtO_2$ is <20 mmHg, potential mechanisms of hypoxia should be explored; see **Figure 1** [44, 45]:

- **Ischemic hypoxia:** ensure adequate CBF using TCD, and maintain optimal CPP.

- **Low oxygen extraction hypoxia:** exclude hypoxemia (maintain PaO_2 >65 mmHg), anemia (maintain hemoglobin >7–8 g/dL), and conditions associated with increased hemoglobin–oxygen affinity, such as hypothermia, hypophosphatemia, alkalosis, and hypocapnia. In carbon monoxide poisoning, hemoglobin-binding sites are occupied by CO, preventing effective oxygen transport.

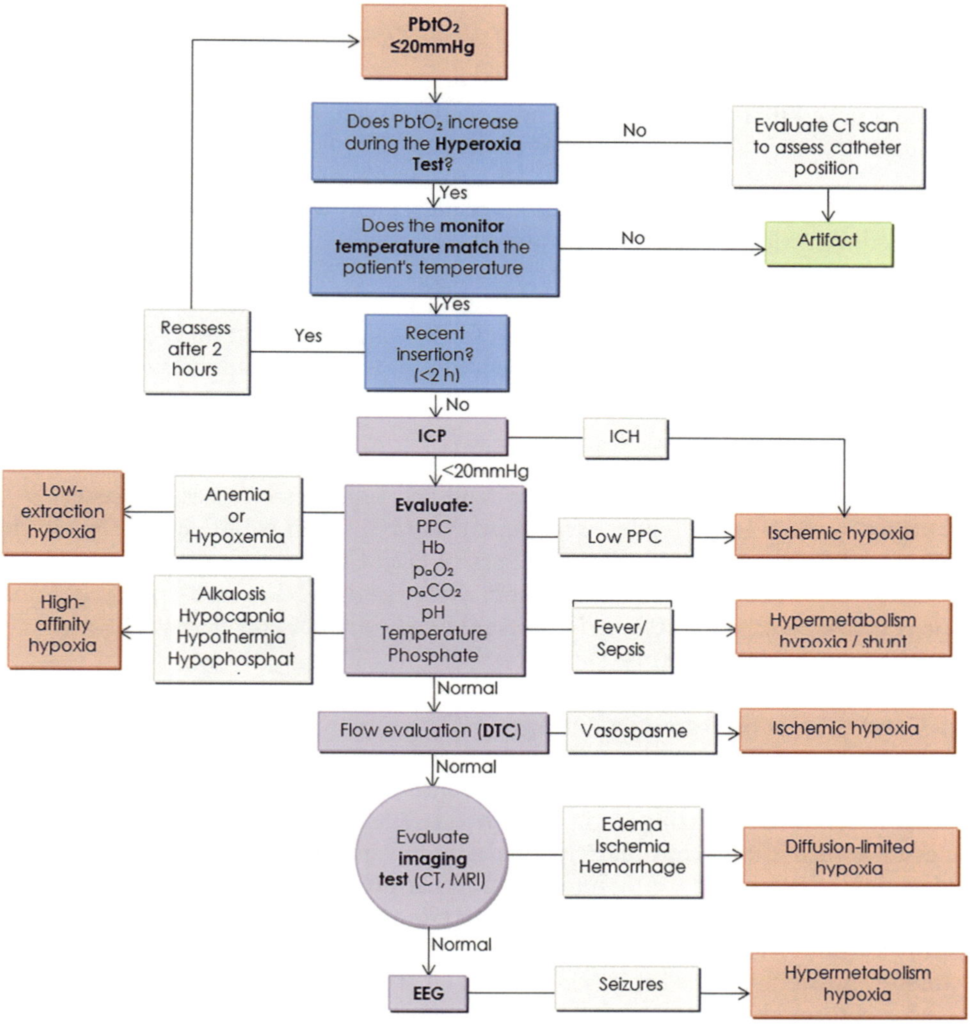

Figure 1.
Differential diagnostic algorithm for low brain tissue oxygenation ($PbtO_2$). *$Pbto_2$: Brain tissue oxygen tension; ICH: Intracranial hypertension; EEG: Electroencephalogram; TCD: Transcranial Doppler; ICP: intracranial pressure; Hb: hemoglobin; $PaCO_2$: arterial carbon dioxide pressure; PaO_2: arterial oxygen pressure; CPP: cerebral perfusion pressure; MRI: magnetic resonance imaging.*

- **Diffusion-limited hypoxia:** situations in which oxygen does not adequately reach the pericellular interstitium due to tissue edema or microthrombosis.

- **Hypermetabolic hypoxia:** exclude fever and epileptic seizures (using EEG), and treat the underlying causes of increased metabolic demand.

A $PbtO_2$ value ≥20 mmHg does not guarantee adequate cellular oxygenation. There are clinical scenarios in which normal or elevated $PbtO_2$ values may still be associated with cellular hypoxia:

- **Extensive cellular ischemia:** when widespread cellular death is present, oxygen is not utilized and may remain normal or elevated in the interstitial space.

- **Mitochondrial dysfunction hypoxia:** due to exogenous toxins (such as cyanide or smoke inhalation containing cyanogenic compounds, frequently produced during plastic combustion) or secondary to TBI [46, 47].

- **Probe malposition:** incorrect catheter placement within cortical regions may produce artifactually elevated values [11].

To assess the reliability of measured $PbtO_2$ values, a hyperoxia challenge test may be performed. This consists of a transient increase in FiO_2 to 1.0 for 2–5 minutes and evaluation of the $PbtO_2$ response. A rapid and significant rise supports correct sensor function; absence of response suggests possible measurement error or probe malfunction [11, 36].

On the other hand, NIRS estimates regional oxygen saturation non-invasively by measuring light absorption through the skull and superficial tissues. However, its utility in aSAH is limited by extracranial contamination, where blood flow in the scalp and skin pigments can distort the signal [12]. Crucially, studies demonstrate that NIRS and $PbtO_2$ monitoring are not interchangeable; NIRS fails to reliably detect values below the critical 20 mmHg threshold and does not correlate well with invasive trajectories over time [13].

2.3 Cerebral metabolism monitoring: CMD

Poor tissue perfusion of the cerebral parenchyma disrupts extracellular homeostasis. Characterizing changes in the extracellular compartment provides relevant information about underlying alterations in cellular metabolism. This principle underlies CMD, a monitoring tool that provides online, semicontinuous analysis of brain tissue biochemistry by sampling the extracellular interstitial fluid (ISF) [23, 48]. The technique utilizes a double-lumen, thin (0.6 mm) dialysis catheter inserted into the brain parenchyma. A sterile perfusate, which mimics CSF, is infused through the inner tube at a constant rate, typically 0.3 μL/min. At the catheter tip, a semipermeable membrane allows molecules to diffuse down their concentration gradient from the ISF into the perfusate (see **Figure 2**). This returning fluid, known as the dialyzate, is collected for biochemical analysis, usually at hourly intervals [10, 49].

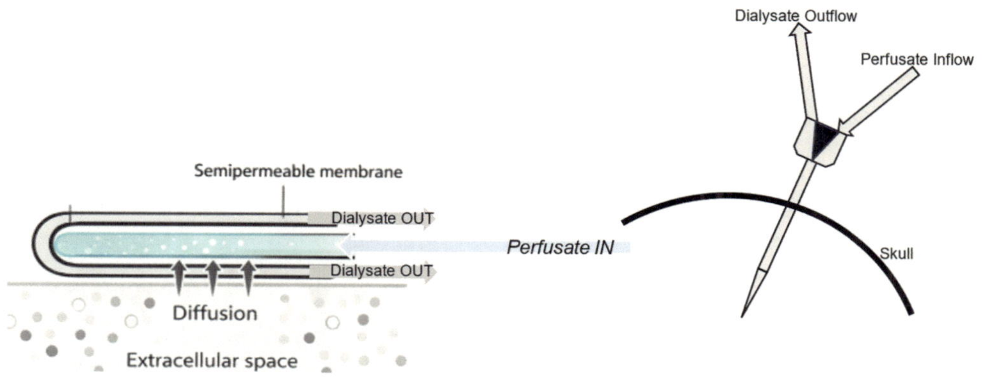

Figure 2.
Detail of cerebral microdialysis catheter functioning. The microdialysis catheter allows continuous sampling of extracellular brain metabolites through a semipermeable membrane, enabling monitoring of cerebral biochemical changes.

In clinical practice, the choice of membrane and location is critical [10, 23, 48, 49]:

- Membrane: A standard 20 kDa molecular weight cut-off is used for energy metabolites, while 100 kDa membranes allow for the recovery of larger molecules, like inflammatory cytokines.

- Location: In aSAH, the probe is ideally placed in the frontal lobe watershed area (ACA–MCA territory) on the side of the maximal blood load or the ruptured aneurysm. However, if a hypoperfused territory is identified on neuroimaging (for example, by CT perfusion), the probe may be placed in the at-risk region.

- Stabilization: Samples collected during the first 1–2 hours post-insertion (the run-in time) are discarded due to microtrauma and local artifacts.

CMD monitoring focuses on markers that reflect the bioenergetic state and structural integrity of the brain [23, 48, 49]:

- **Glucose:** The primary energy substrate. Concentrations <0.8 mmol/L indicate a restricted supply, while levels <0.2 mmol/L represent critical neuroglycopenia.

- **Lactate and Pyruvate:** These reflect the glycolytic pathway. While absolute lactate >4.0 mmol/L indicates metabolic stress, it must be interpreted alongside the Lactate/Pyruvate ratio.

- **Lactate/Pyruvate Ratio (LPR)**: The most robust indicator of the cellular redox status. An LPR >25 is abnormal, and >40 marks a critical metabolic energy crisis.

- **Glutamate**: An excitatory neurotransmitter. Elevations reflect energy failure and the disruption of astrocytic reuptake, contributing to excitotoxicity.

- **Glycerol**: A byproduct of cell membrane phospholipid degradation and a marker of the loss of cellular structural integrity.

CMD may help distinguish different metabolic patterns of energy failure based on the relationship between LPR, pyruvate, and glucose [23, 48–50]:

(1) **Classic Ischemia (Type 1 Energy Crisis):** Defined by an elevated LPR (>40), accompanied by low pyruvate and low glucose, indicating a failure of both substrate and oxygen delivery.

(2) **Mitochondrial Dysfunction (Type 2/Non-Ischemic Energy Crisis):** Characterized by an elevated LPR, but with normal or high pyruvate levels, suggesting a failure of oxygen utilization at the mitochondrial level.

CMD changes may precede clinical or radiological signs of DCI [10]. Metabolic derangements – specifically a pattern of increasing LPR and simultaneously decreasing glucose – can precede clinical or radiological signs of DCI by 16 to 18 hours [20]. This allows for a proactive "intention-to-treat" strategy, where clinicians may increase CPP or optimize volemia to reverse incipient cerebral distress before permanent tissue infarction occurs. Recent studies show that adherence to CMD-based protocols can lead to a significant reduction in the incidence of DCI-related infarcts [10, 23].

For example, a rise in ICP accompanied by a drop in $PbtO_2$ <20 mmHg and an elevated CMD LPR >to 40 provides a strong physiological rationale for escalation to tier-two interventions, such as induced hypertension or CSF drainage, with the goal of reversing impending cellular energy failure. Moving from isolated threshold-based triggers to an integrated, physiology-guided strategy allows therapies to be better tailored to the dynamic metabolic demands of the injured brain [50].

2.4 Electrical activity monitoring tools: Electroencephalography

Continuous EEG (cEEG) is widely used in the multimodal monitoring of aSAH due to the high prevalence of "silent" electrical events. NCS and periodic discharges (PDs) occur in approximately 61.8% of patients with severe acute brain injury, including subarachnoid hemorrhage (SAH) [25]. Because these events lack motor accompaniments, they are frequently undetectable by bedside clinical examination alone [51].

In the context of aSAH, cEEG may reflect established injury and can, in some cases, precede clinical deterioration related to DCI. In high-grade SAH patients, cEEG can reveal new-onset ischemic neuronal deficits before they become radiologically apparent. A critical phenomenon detected via invasive monitoring is cortical spreading depolarization (CSD), which occurs in over half of severe SAH patients. CSDs represent massive waves of cellular depolarization that trigger a "vicious cycle" of glucose depletion and parenchymal hypoperfusion, directly contributing to the expansion of ischemic zones and subsequent infarction [22, 48].

Quantitative EEG Trends: To facilitate real-time clinical integration, quantitative EEG (qEEG) software is increasingly utilized at the bedside. These systems use automated computer algorithms to process raw EEG data into visual trends, such as compressed spectral arrays, which allow clinicians to identify seizure activity and ischemic shifts more rapidly [22, 25].

Clinical Integration and Limitations: The integration of cEEG into the neuro-intensive care unit workflow remains challenging due to several factors:

- Skill requirements: Accurate interpretation of raw EEG data requires specialized training and is highly labor-intensive.

- Invasiveness: While scalp EEG is noninvasive, it often misses deep-seated seizures and cannot detect cortical spreading depolarizations, which require the placement of subdural strip or depth electrodes [22].

- Technological gaps: Models focused on intracranial compliance often omit EEG, which is a limitation because metabolic crises and CSDs can occur unnoticed by pressure or oxygen monitors alone [27].

NCS may occur both during the early phase of acute brain injury after hemorrhage and as a manifestation of hypoperfusion related to DCI. All patients presenting with a reduced level of consciousness should undergo EEG evaluation at admission to exclude NCS as a potential cause of coma. In addition, abrupt neurological deterioration, evidence of a metabolic crisis detected by CMD, or an otherwise unexplained decrease in $PbtO_2$ should be considered warning signs suggestive of epileptic activity.

3. Toward proactive and personalized care in aSAH

The integration of different neuromonitoring modalities into a multimodal monitoring paradigm has transformed the management of aSAH from reactive treatment to a proactive, "intention-to-treat" strategy [22, 27]. This shift is based on the understanding that no single variable can capture the full complexity of post-hemorrhage brain injury.

In clinical practice, trends are often more informative than isolated values [23]. ICP trajectory monitoring allows for the early detection of "incipient cerebral distress" [20]. Similarly, daily TCD examinations showing flow velocity increases of >50 cm/s over 24 hours are highly predictive of symptomatic vasospasm and impending ischemia. By integrating these trajectories, protocols can trigger interventions – such as CSF drainage or individualized CPP optimization – hours before irreversible tissue damage or clinical deterioration manifests [39].

We propose a stepwise (step-up) multimodal monitoring framework for patients with aSAH, in which monitoring intensity is progressively escalated according to clinical severity, risk profile, and early clinical course:

- **Tier 0 (Clinical and noninvasive baseline monitoring):** Clinical neurological examination, daily TCD, and EEG when the neurological status is not congruent with the expected clinical course.

- **Tier 1 (Standard):** ICP and CPP may be considered as a first-line strategy in patients with GCS >9 or in sedated patients with a high intracranial blood burden (modified Fisher grade 3–4), particularly to detect and treat episodes of ICH during the EBI phase of aSAH [52].

- **Tier 2 (Advanced oxygen-guided monitoring):** The addition of $PbtO_2$ to ICP monitoring is recommended in patients with an unfavorable early course who were initially monitored with ICP alone (e.g., those with recurrent ICH or persistent coma), as well as in the most severe cases (WFNS/Hunt–Hess grade V) or in patients with systemic complications such as hypoxemia or shock [34, 39].

- **Tier 3 (Comprehensive metabolic monitoring):** CMD is generally implemented as a third-tier monitoring modality in high-risk patients or in those with persistent metabolic crises despite optimized ICP and brain oxygenation. Despite its high sensitivity, its focal nature must be recognized, as CMD may fail to detect ischemic events occurring in the contralateral hemisphere or in regions located more than approximately 4 cm from the probe. CMD data should, therefore, always be interpreted within a multimodal framework, integrated with ICP, $PbtO_2$, and neuroimaging findings to obtain a comprehensive assessment of cerebral pathophysiology.

Multimodal monitoring may facilitate earlier recognition of secondary insults and support timely intervention, allowing physicians to deliver truly personalized care that targets the cellular and metabolic needs of the injured brain [27, 50].

Author details

Manel Santafè Colomina
Hospital Universitari Bellvitge, Intensive Care Unit, L'Hospitalet de Llobregat, Barcelona, Spain

*Address all correspondence to: santafe.manel@gmail.com

References

[1] Hoh BL, Ko NU, Amin-Hanjani S, Chou S-Y, Cruz-Flores S, Dangayach NS, et al. Guideline for the management of patients with aneurysmal subarachnoid hemorrhage: A guideline from the american heart association/american stroke association. Stroke. 2023;**54**:e314–e70

[2] Mayer SA, Kreiter KT, Copeland D, Bernardini GL, Bates JE, Peery S, et al. Global and domain-specific cognitive impairment and outcome after subarachnoid hemorrhage. Neurology. 2002;**59**:1750–1758

[3] Ingall T, Asplund K, Mähönen M, Bonita R. A multinational comparison of subarachnoid hemorrhage epidemiology in the WHO MONICA stroke study. Stroke. 2000;**31**(5):1054–1061

[4] Moussouttas M, Lai EW, Dombrowski K, Huynh TT, Khoury J, Carmona G, et al. CSF catecholamine profile in subarachnoid hemorrhage patients with neurogenic cardiomyopathy. Neurocritical Care. 2011;**14**:401–406

[5] Moussouttas M, Mearns E, Walters A, DeCaro M. Plasma catecholamine profile of subarachnoid hemorrhage patients with neurogenic cardiomyopathy. Cerebrovascular Diseases Extra. 2015;**5**(2):57–67

[6] Salem R, Vallée F, Dépret F, Callebert J, Maurice JP, Marty P, et al. Subarachnoid hemorrhage induces an early and reversible cardiac injury associated with catecholamine release: One-week follow-up study. Critical Care (London, England). 2014;**18**:558

[7] Deenen S, Ramnarain D, Pouwels S. Subarachnoidal hemorrhage related cardiomyopathy: An overview of Tako-Tsubo cardiomyopathy and related cardiac syndromes. Expert Review of Cardiovascular Therapy. 2022;**20**(9):733–745

[8] Ramnarain D, Deenen S, Boerboom L, van Oers J, Pouwels S. Neurogenic stunned myocardium: A narrative review of cardiopulmonary failure related to subarachnoidal haemorrhage. Cureus. 2025;**17**(9): e93133

[9] Muroi C, Keller M, Pangalu A, Fortunati M, Yonekawa Y, Keller E. Neurogenic pulmonary edema in patients with subarachnoid hemorrhage. Journal of Neurosurgical Anesthesiology. 2008;**20**(3):188–192

[10] Winberg J, Holm I, Cederberg D, Rundgren M, Kronvall E, Marklund N. Cerebral microdialysis-based interventions targeting delayed cerebral ischemia following aneurysmal subarachnoid hemorrhage. Neurocritical Care. 2022;**37**(1):255–266

[11] Dings J, Meixensberger J, Jäger A, Roosen K. Clinical experience with 118 brain tissue oxygen partial pressure catheter probes. Neurosurgery. 1998;**43**(5):1082–1095

[12] Robba C, Battaglini D, Rasulo F, Lobo FA, Matta B. The importance of monitoring cerebral oxygenation in non brain injured patients. Journal of Clinical Monitoring and Computing. 2023;**37**(4):943–949

[13] de Courson H, Proust-Lima C, Tuaz E, Georges D, Verchère E, Biais M.

Relationship between brain tissue oxygen and near-infrared spectroscopy in patients with nontraumatic subarachnoid hemorrhage. Neurocritical Care. 2022;**37**(3):620–628

[14] Neulen A, Meyer S, Kramer A, Pantel T, Kosterhon M, Kunzelmann S, et al. Large vessel vasospasm is not associated with cerebral cortical hypoperfusion in a murine model of subarachnoid hemorrhage. Translational Stroke Research. 2018;**10**:319–326

[15] Clarke JV, Suggs JM, Diwan D, Lee JV, Lipsey K, Vellimana AK, et al. Microvascular platelet aggregation and thrombosis after subarachnoid hemorrhage: A review and synthesis. Journal of Cerebral Blood Flow and Metabolism: Official Journal of the International Society of Cerebral Blood Flow and Metabolism. 2020;**40**:1565–1575

[16] Schallner N, Pandit R, LeBlanc R, AJ 3T, CS O, BS Z, et al. Microglia regulate blood clearance in subarachnoid hemorrhage by heme oxygenase-1. The Journal of Clinical Investigation. 2015;**125**: 2609–2625

[17] Geraghty JR, Davis JL, Testai FD. Neuroinflammation and microvascular dysfunction after experimental subarachnoid hemorrhage: Emerging components of early brain injury related to outcome. Neurocritical Care. 2019;**31**(2):373–389

[18] Hartings JA, Wilson JA, Look AC, Vagal A, Shutter LA, Dreier JP, et al. Full-band electrocorticography of spreading depolarizations in patients with aneurysmal subarachnoid hemorrhage. Acta Neurochirurgica Supplement. 2013;**115**:131–141

[19] Bosche B, Graf R, Ernestus RI, Dohmen C, Reithmeier T, Brinker G, et al. Recurrent spreading depolarizations after subarachnoid hemorrhage decreases oxygen availability in human cerebral cortex. Annals of Neurology. 2010;**67**:607–617

[20] Patet C, Quintard H, Zerlauth JB, Maibach T, Carteron L, Suys T, et al. Bedside cerebral microdialysis monitoring of delayed cerebral hypoperfusion in comatose patients with poor grade aneurysmal subarachnoid haemorrhage. Journal of Neurology, Neurosurgery & Psychiatry. 2017;**88**:332–338

[21] Santafé Colomina M, Arikan Abelló F, Sánchez Corral A, Ferrer Roca R. Optimization of the neurosurgical patient in Intensive Care. Medicina Intensiva. 2019;**43**(8):489–496

[22] Smith M. Multimodality Neuromonitoring in Adult Traumatic Brain Injury: A Narrative Review. Anesthesiology. 2018;**128**(2):401–415

[23] Hutchinson PJ, Jalloh I, Helmy A, Carpenter KL, Rostami E, Bellander BM, et al. Consensus statement from the 2014 International Microdialysis Forum. Intensive Care Medicine. 2015;**41**:1517–1528

[24] Blanco P, Abdo-Cuza A. Transcranial Doppler ultrasound in neurocritical care. Journal of Ultrasound. 2018;**21**(1):1–16

[25] Vespa P, Tubi M, Claassen J, Buitrago-Blanco M, McArthur D, Velazquez AG, et al. Metabolic crisis occurs with seizures and periodic discharges after brain trauma. Annals of Neurology. 2016;**79**:579–590

[26] Lindegaard KF, Nornes H, Bakke SJ, Sorteberg W, Nakstad P.

Cerebral vasospasm after subarachnoid haemorrhage investigated by means of transcranial Doppler ultrasound. Acta Neurochirurgica Supplementum. 1988;**42**:81–84

[27] Godoy DA, Brasil S, Iaccarino C, Paiva W, Rubiano AM. The intracranial compartmental syndrome: A proposed model for acute brain injury monitoring and management. Critical Care (London, England). 2023;**27**(1):137

[28] Rasulo FA, Calza S, Robba C, Taccone FS, Biasucci DG, Badenes R, et al. Transcranial Doppler as a screening test to exclude intracranial hypertension in brain-injured patients: The IMPRESSIT-2 prospective multicenter international study. Critical Care (London, England). 2022;**26**:110

[29] Kumar G, Shahripour RB, Harrigan MR. Vasospasm on transcranial Doppler is predictive of delayed cerebral ischemia in aneurysmal subarachnoid hemorrhage: A systematic review and meta-analysis. Journal of Neurosurgery. 2016;**124**(5):1257–1264

[30] Al-Nusair L, Kouli O, Ansari S, Carey K, Farhad I, Ali A, Nunn A, Lakhani S, Hannan CJ, Yousaf J. Impact of transcranial Doppler on the management of delayed cerebral ischemia: A propensity-matched analysis. Journal of Neurosurgery. 2026;1–10. DOI: 10.3171/2025.9

[31] Addis A, Baggiani M, Citerio G. Intracranial pressure monitoring and management in aneurysmal subarachnoid hemorrhage. Neurocritical Care. 2023;**39**(1):59–69

[32] Svedung Wettervik T, Hånell A, Howells T, Ronne-Engström E, Lewén A, Enblad P. Intracranial pressure- and cerebral perfusion pressure threshold-insults in relation to cerebral energy metabolism in aneurysmal subarachnoid hemorrhage. Acta Neurochirurgica. 2022;**164**(4):1001–1014

[33] Rapp A, Kobeissi H, Fahim DK. Updated review of the management of and guidelines for traumatic brain injury. Journal of Clinical Medicine. 2025;**14**(19):6796

[34] Robba C, Taccone FS, Citerio G. Monitoring cerebral oxygenation in acute brain-injured patients. Intensive Care Medicine. 2022;**48**(10):1463–1466

[35] Megjhani M, Weiss M, Ford J, Terilli K, Kastenholz N, Nametz D, et al. Optimal cerebral perfusion pressure and brain tissue oxygen in aneurysmal subarachnoid hemorrhage. Stroke. 2023;**54**:189–197

[36] Marín-Caballos AJ, Murillo-Cabezas F, Domínguez-Roldan JM, Leal-Noval SR, Rincón-Ferrari MD, Muñoz-Sánchez MA. Monitoring of tissue oxygen pressure (PtiO2) in cerebral hypoxia: Diagnostic and therapeutic approach. Medicina Intensiva. 2008;**32**(2):81–90

[37] Pennings FA, Schuurman PR, van den Munckhof P, Bouma GJ. Brain tissue oxygen pressure monitoring in awake patients during functional neurosurgery: The assessment of normal values. Journal of Neurotrauma. 2008;**25**(10):1173–1177

[38] Roh D, Park S. Brain multimodality monitoring: Updated perspectives. Current Neurology and Neuroscience Reports. 2016;**16**(6):56

[39] Le Roux P, Menon DK, Citerio G, Vespa P, Bader MK, Brophy GM, et al.

Consensus summary statement of the International Multidisciplinary Consensus Conference on Multimodality Monitoring in Neurocritical Care: A statement for healthcare professionals from the Neurocritical Care Society and the European Society of Intensive Care Medicine. Intensive Care Medicine. 2014;**40**:1189–1209

[40] Dengler J, Frenzel C, Vajkoczy P, Wolf S, Horn P. Cerebral tissue oxygenation measured by two different probes: Challenges and interpretation. Intensive Care Medicine. 2011;**37**(11):1809–1815

[41] Meixensberger J, Dings J, Kuhnigk H, Roosen K. Studies of tissue PO2 in normal and pathological human brain cortex. Acta Neurochirurgica Supplementum. 1993;**59**:58–63

[42] Clark Jr LC, Kaplan S, Matthews EC, Edwards FK, Helmsworth JA. Monitor and control of blood oxygen tension and pH during total body perfusion. The Journal of Thoracic Surgery. 1958;**36**(4): 488–496

[43] Huschak G, Hoell T, Hohaus C, Kern C, Minkus Y, Meisel HJ. Clinical evaluation of a new multiparameter neuromonitoring device: Measurement of brain tissue oxygen, brain temperature, and intracranial pressure. Journal of Neurosurgical Anesthesiology. 2009;**21**(2):155–160

[44] Siggaard-Andersen O, Ulrich A, Gøthgen IH. Classes of tissue hypoxia. Acta Anaesthesiologica Scandinavica Supplementum. 1995;**107**:137–142

[45] Nathan AT, Singer M. The oxygen trail: Tissue oxygenation. British Medical Bulletin. 1999;**55** (1):96–108

[46] Hiebert JB, Shen Q, Thimmesch AR, Pierce JD. Traumatic brain injury and mitochondrial dysfunction. The American Journal of the Medical Sciences. 2015;**350** (2):132–138

[47] Dobrachinski F, da Rosa Gerbatin R, Sartori G, Ferreira Marques N, Zemolin AP, Almeida Silva LF, et al. Regulation of mitochondrial function and glutamatergic system are the target of guanosine effect in traumatic brain injury. Journal of Neurotrauma. 2017;**34**:1318–1328

[48] Spencer P, Jiang Y, Liu N, Han J, Li Y, Vodovoz S, et al. Update: Microdialysis for monitoring cerebral metabolic dysfunction after subarachnoid hemorrhage. Journal of Clinical Medicine. 2020;**10**:100

[49] de Lima Oliveira M, Kairalla AC, Fonoff ET, Martinez RC, Teixeira MJ, Bor-Seng-Shu E. Cerebral microdialysis in traumatic brain injury and subarachnoid hemorrhage: State of the art. Neurocritical Care. 2014;**21** (1):152–162

[50] Carteron L, Bouzat P, Oddo M. Cerebral microdialysis monitoring to improve individualized neurointensive care therapy: An update of recent clinical data. Frontiers in Neurology. 2017;**8**:601

[51] Asaithambi G, Martins SL. Role of electroencephalography in the management of aneurysmal

subarachnoid hemorrhage: National Inpatient Sample, 2012-2017. Journal of Neurosurgical Anesthesiology. 2025;**38**:172–175

[52] Chesnut R, Aguilera S, Buki A, Bulger E, Citerio G, Cooper DJ, et al. A management algorithm for adult patients with both brain oxygen and intracranial pressure monitoring: The Seattle International Severe Traumatic Brain Injury Consensus Conference (SIBICC). Intensive Care Medicine. 2020;**46**:919–929

Chapter 3

Management of Neuro-Oncologic Emergencies

Alicia Dupre and Andrew Treister

Abstract

Patients with primary or secondary neoplastic disease of the central nervous system are at an intrinsically high risk of developing neurologic emergencies related to their disease. Appropriate and expedited management of these emergencies can have profound implications on mortality, duration of life expectancy, and quality of life. We present an overview of emergencies that may be encountered by neuro-oncologists, neurologists, neurointensivists, or neurosurgeons who treat patients with neuro-oncologic disease. Many such crises can erupt as a direct result of the tumor itself, whereby the mass effect can lead to symptomatic cerebral edema, elevated intracranial pressure, and herniation syndromes. The physiology and anatomic considerations are reviewed within the context of approaches to treatment. Tumors that disrupt the regular flow of cerebrospinal fluid (CSF) through the ventricular system can also cause obstructive hydrocephalus and may require CSF diversion to prevent catastrophic complications. Seizures and status epilepticus are common among those with brain tumors due to multiple factors that can influence cortical irritability. A multi-step appraoch to the management of seizures and status epilepticus in those with brain tumors is described. Risk factors and treatment methods for patients with cancer who develop arterial and venous thrombotic disease is addressed as well. Risk factors and characteristics of patients who develop spinal cord compression related to neoplastic disease are presented with an emphasis on early interventions to prevent functional disabilities. Finally, treatment complications related to the emerging therapy, immune effector cell-associated neurotoxicity syndrome (ICANS) is discussed in detail. A solid understanding and anticipation of the complications associated with neurologic cancer care is paramount to achieving the best outcomes for these patients.

Keywords: brain tumors, cerebral edema, herniation, obstructive hydrocephalus, status epilepticus, spinal cord compression, ICANS

Literature search

Database searches were carried out primarily using PubMed in January and February 2026. An initial search for "Neurooncologic Emergencies" yielded 35 results and was used to identify key topics for discussion and their primary sources. Subsequent searches, intending to add more recent literature, included "brain tumor

AND cerebral edema AND physiology" (2,324 results, "brain tumor AND cerebral edema AND treatment (4,300 results), "brain tumor AND hydrocephalus" (5,274 results), "brain tumor AND status epilepticus" (432 results), "cancer AND venous sinus thrombosis" (549 results), "ICANS treatment" (1,261 results). Searches were further refined by prioritizing newer studies with broader applicability to the chosen topics. Small case series and articles detailing expert opinion were included where robust randomized controlled trials had not been performed.

1. Introduction

By 2045, there will be an estimated 32.6 million new cancer diagnoses worldwide: an increase from 20.0 million in 2022. Of these, roughly 474,000 are thought to be primary brain or central nervous system (CNS) cancers (an increase from 322,000 in 2022), with an estimated mortality of 383,000 [1]. Currently, cancers of all types are a leading cause of death worldwide, accounting for nearly 1 in 6 deaths (roughly 10 million annually), with 3 in 10 deaths being premature (ages 30–69 years), second only to cardiovascular disease [2, 3]. While overall the incidence of new primary brain and CNS cancer diagnoses has gradually but slowly declined since a peak in 1999 at 6.7 per 100,000 cases to 6.1 per 100,000 cases in 2025, the rate of death has remained relatively steady from 4.6 per 100,000 cases to 4.4 per 100,000 cases in 2025 [4]. Although more rare than other cancer types, the mortality from brain and other nervous system cancers is high, with an estimated five-year relative survival of 33.0% when evaluating survival data taken from 2015–2021 [4].

In addition to primary brain and CNS tumors, brain metastases represent the most common intracranial neoplasms in adult patients, although their prevalence is often difficult to capture in cancer registries, with most data coming from historical study estimates and autopsies [5, 6]. Per a report by Taboret et al. in 2012 examining trends in the epidemiology of brain metastases, the estimated number of new cases of brain metastases in the USA per year is roughly 170,000, with a global prevalence of 8.5–9.6%, although this is likely underestimated [6]. This has been increasing over the preceding 20 years and is higher than the annual incidence of primary brain tumors [6, 7]. Potentially, this is due to improvements in survivability timelines for systemic cancers that spread to the brain, as treatment options have increased. The main systemic solid organ cancers that metastasize to the brain include lung – especially small cell lung cancers – renal, melanoma, and breast (in order of prevalence), while those that prefer to spread to the spine are lung, breast, prostate, renal, gastrointestinal, thyroid, non-Hodgkin's lymphoma, and multiple myeloma [7].

Although the leading cause of death in patients with brain metastases remains the advancement and complications of their systemic disease, there are several acute and critical complications that can arise from the presence of intracranial lesions, both primary and metastatic [7]. With the high disease burden of cancers throughout the USA and the world, the recognition and treatment of these neuro-oncologic emergencies are essential for improving patient morbidity and mortality. These emergent neurologic complications can result from anatomic or physiologic features of the tumors themselves, or from treatment modalities.

2. Tumor-associated emergencies

Several neurologic emergencies can be tied to either direct tumor-mediated factors or the anatomical location of intracranial or spinal lesions and impact both morbidity and mortality in cancer patients. The most common malignant primary brain tumor is glioblastoma multiforme (GBM) (WHO grade IV), representing 14.2% of all brain tumors and more than half of newly diagnosed malignant intracranial tumors, while meningiomas make up the most common nonmalignant brain lesions at 55.4% [7]. Both malignant and nonmalignant lesions can mediate neurologic emergencies directly from their anatomic positioning, as well as indirectly from factors such as coagulopathies, antibodies, and enhanced autoimmunity. Often, the direct tumor-mediated neuro-oncologic emergencies are rapidly life-threatening and require urgent identification and intervention to stave off brain destruction and death.

2.1 Intracranial hypertension and cerebral edema

As first proposed in the 1820s, the Monro-Kellie doctrine emphasizes the need for balance between brain parenchyma, cerebrospinal fluid (CSF), and blood to maintain intracranial homeostasis [8]. Any increase in one must come at the expense of another due to the noncompliant nature of the adult skull. Within this fixed volume of 1,400–1,700 ml, around 80% is brain parenchyma, 10% is CSF, and 10% is blood [9]. In brain tumors, both metastatic and primary, there is an unexpected increase in brain parenchyma beyond the allotted 80%, altering this delicate balance, and while some CSF can circulate out via the arachnoid granulations and vasoconstriction can lessen blood flow, there is a limit to the safety of these compensatory mechanisms [10]. When intracranial lesions are slow-growing, the brain can accommodate an increase in tissue relatively well without symptoms, until a critical threshold is reached, which can vary by patient and disease [7]. Intracranial hypertension is the life-threatening consequence of surpassing this threshold, resulting in hydrocephalus, herniation, and death if not addressed.

Defined as an intracranial pressure (ICP) greater than 20–25 mmHg, intracranial hypertension is one of the most common emergencies in neuro-oncology [11]. Prior studies have shown a proportional relationship between ICP and mortality, with high ICP between 20 and 30 mmHg being associated with increased mortality [12]. Additionally, there is an inverse relationship between cerebral perfusion pressure (CPP) and mortality, with higher mortality observed when CPP is less than 55 mmHg [12]. In neuro-oncology, several factors can lead to elevated ICP, including cerebral edema, mass effect from the tumor itself, and obstruction of the ventricular system, either due to tumor placement or leptomeningeal spread [11]. Regardless of etiology, because of the relationship between ICP and CPP, where CPP is determined by mean arterial pressure (MAP) minus ICP (CPP = MAP – ICP), intracranial hypertension leads to impaired tissue oxygenation, abnormal cerebral metabolism, and a decline in cerebral blood flow, thus causing secondary brain injury and perpetuating a dangerous spiral [11, 13].

Depending on the acuity of ICP elevation and patient characteristics, the early signs and symptoms of intracranial hypertension range from headache – ranging from tension-type to severe and accompanied by transient neurologic deficits and plateau waves of heightened ICPs over 40 mmHg – confusion, and nausea to focal neurologic symptoms, blurred vision, and altered levels of consciousness, including coma when severe [9, 11]. Half of patients will present initially with headache, even prior to a known cancer diagnosis, which they describe as throbbing and pressure-like, worse in the morning, and not helped by analgesics [10]. In one-third of cases, there will be loss of consciousness that often signals impending herniation [10]. With this, patients can present with abnormal breathing consistent with a Cheyne-Stokes or Central Neurogenic Hyperventilation pattern; hiccups or yawning; and Cushing's Triad of hypertension, bradycardia, and irregular respirations, depending on what areas of the brain are herniating [10].

Classically, in the setting of both metastatic and primary brain tumors, high ICP arises from vasogenic edema surrounding the tumor, which accumulates at an average rate of 17–78 ml/day [14]; direct mass effect from the tumor itself; or obstructive hydrocephalus when the ventricular system is impacted by tumor location or spread [10]. Imaging findings in intracranial hypertension will depend on the cause of the elevated ICP. A noncontrasted head CT should be obtained once immediate management of high ICP has been undertaken to rule out causes such as hemorrhage, followed by contrast-enhanced MRIs to look for edematous changes, lesion expansion, and sequelae of obstructive hydrocephalus or meningeal spread of disease [9, 10].

Amongst the potential causes of cerebral edema, vasogenic edema is the most common underlying etiology in intracranial malignancies and results from a multifactorial disruption of the blood–brain barrier that allows plasma and proteins into the interstitial space. Unlike cytotoxic edema seen in ischemic states, vasogenic edema seen in intracranial malignancy involves expansion of the extracellular – rather than intracellular – compartment in either the CSF or interstitial fluid [14]. Other instances of extracellular edema include hydrocephalic edema from CSF obstruction, osmotic edema from an altered gradient (water intoxication, acute hyponatremia), and stasis, which can be seen from tumor compression of venous drainage, can also generate edema via expansion of the extracellular space [14].

On the histological scale, vasogenic edema in malignancies stems from several factors. The tight junctions of the blood–brain barrier become disrupted and dysfunctional, allowing for a loss of homogeneity in the basement membranes [7]. With this, there is an increase in pinocytic activity, as well as upregulation of vascular endothelial growth factor (VEGF), a vasoactive proangiogenic molecule that occurs in both malignant and benign primary and metastatic CNS tumors. While expressed in normal brain tissue, VEGF has increased expression following brain injuries [7, 15]. Higher VEGF activity in malignancies creates structurally normal but functionally abnormal blood vessels, which, in turn, contribute to leakage of the blood–brain barrier and the accumulation of interstitial water and sodium outside of the intravascular space, as VEGF activity increases blood–brain barrier permeability to molecules of at least moderate size [15]. In addition to the generation of numerous but permeable blood vessels, VEGF contributes to the synthesis and release of nitric oxide, which causes additional blood–brain barrier permeability and vasodilation of cerebral arterioles [14, 15]. Additional studies have shown that VEGF function during brain injury leads to phosphorylation of occludin – a protein found in tight junctions of normal brain – thus impairing its function and resulting in the opening of tight junctions and further edema accumulation, in conjunction with inducing fenestrations on otherwise nonfenestrated blood–brain barrier endothelium and increasing capillary leak [14].

Importantly, increased VEGF expression has been seen in certain meningiomas, including transitional meningiomas and meningotheliomatous meningiomas, as well as high-grade gliomas (GBM), where its expression is largely controlled by hypoxia and restricted to perinecrotic cells [10, 14]. Taken all together, VEGF plays a key role in the generation of the vasogenic edema seen in intracranial malignancy and was thus felt to be a key target for treatment to lessen the morbidity and mortality of this neoplastic complication with the use of steroids, osmotherapy, and targeted monoclonal antibody agents specifically against VEGF.

2.2 Hydrocephalus

Elevated ICP can both gradually and acutely arise from vasogenic edema in the setting of cerebral neoplasms, but it can also immediately occur due to hydrocephalus. Brain tumors themselves – metastatic and primary – can cause either obstructive or communicating hydrocephalus. Depending on anatomic location, malignancies can hinder the flow dynamics of CSF, creating a stoppage or dam-like effect and causing build-up and backflow. Structurally, the most common causes of obstructive hydrocephalus are posterior fossa tumors due to their location and ability to compress the fourth ventricle (**Figure 1**), supratentorial masses that generate enough mass effect and midline shift to efface the lateral and third ventricles, and intraventricular tumors themselves [7]. Of these, posterior fossa tumors continue to carry a risk of obstructive hydrocephalus even postoperatively, with patients requiring readmission for subsequent CSF flow diversion procedures upwards of seven times more frequently than other tumor locales [7]. This risk is heightened when posterior fossa lesions have the following findings: intraventricular location, solid composition, intraventricular blood on imaging, high-grade tumor, or the presence of metastases. For this reason, when posterior fossa masses are removed, intraprocedural CSF diversions are often performed, including shunt placements or endoscopic third ventriculostomies [7].

A solid anatomic obstruction to CSF is not the sole cause of hydrocephalus in cancer patients. Leptomeningeal carcinomatosis – also called neoplastic meningitis – is a rare but serious complication of intracranial and systemic malignancies that

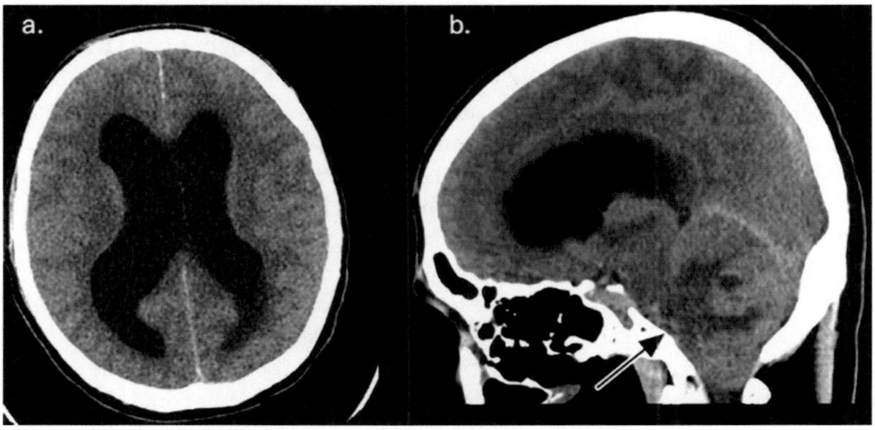

Figure 1.
Obstructive hydrocephalus in the setting of posterior fossa medulloblastoma (a), as noted by hypodensity in cerebellum causing impingement on the fourth ventricle (black arrow) (b).

arises from the spread of cancerous cells to the pia and arachnoid mater [10]. Seeding of cancer cells into the CSF and subsequently to the meningeal layers can occur through hematogenous spread via arterial dissemination or Batson's plexus, spread from adjacent structures such as the choroid plexus or subarachnoid space, or retrograde invasion from perivascular spaces [10]. Regardless, due to obstruction of CSF absorption and flow dynamics in this complication, over two-thirds of patients will develop hydrocephalus, with symptoms often present despite the absence of obstruction on basic imaging [7]. These symptoms are similar to those of other causes of elevated ICP, including headache, alterations of consciousness, vision changes, and nausea, but can also include spinal symptoms and multiple cranial neuropathies due to the infestation of the meningeal covering of each with cancer cells [10].

The presence of leptomeningeal carcinomatosis is observed in advanced disease, indicating a median survival of four to six weeks once detected if untreated, and extending to four to eight months if treatment is pursued [7, 10]. Prognosis increases slightly if it originates from a primary hematologic malignancy (5–15% of cases) compared to solid tumors (1–5% of cases) or primary brain tumors (1–2% of cases) [10]. Diagnosis requires a lumbar puncture to identify cancerous cells in the CSF, although contrast-enhanced MRIs can reveal nodular areas of enhancement in the sulci, basal cisterns, and cranial/spinal nerves that may suggest leptomeningeal carcinomatosis [10]. Aggressive treatment should be pursued in patients starting from a good functional baseline with low tumor burden overall and a relative lack of other neurologic deficits, who are expected to survive longer than three months. This can include radiotherapy to improve CSF flow, systemic or intrathecal chemotherapy, or palliative ventriculoperitoneal shunts (VPSs) for CSF flow diversion [7, 10]. Of note, in randomized controlled trials, there has been some modest benefit from intrathecal chemotherapy for the management of leptomeningeal carcinomatosis, but median survival remains dismal [10].

2.3 Herniation

The ultimate consequence of intracranial hypertension, and the most common secondary cause of death from CNS tumor progression, is herniation. Among those who succumb to their disease, roughly 73% of neuro-oncology patients will pass away from brain herniation [11]. Herniation, or the malplacement of brain parenchyma leading to tissue death, results from space-occupying lesions and elevated ICP due to the inflexibility of the adult skull. Unlike obstructive hydrocephalus, the most common causes of brain herniation are supratentorial masses, which can cause subfalcine, central, downward, and uncal herniations (**Figure 2**) while infratentorial masses are more prone to upward herniation and tonsillar herniations (**Table 1**) [7]. Diagnosis of herniation is often made in the immediate period by critical exam changes and can be confirmed on imaging.

Symptomatology will vary depending on which components of the brain parenchyma are shifted out of place (**Table 1**); however, some common presentations throughout most herniation syndromes include headache – seen in 50% of patients – decreased level of consciousness resulting in coma due to reticular activating system pathway impaction, pupillary changes, cardiorespiratory compromise, and Cushing's triad, which manifests as hypertension, bradycardia, and abnormal respirations or apnea [11]. Cushing's triad is a late finding that

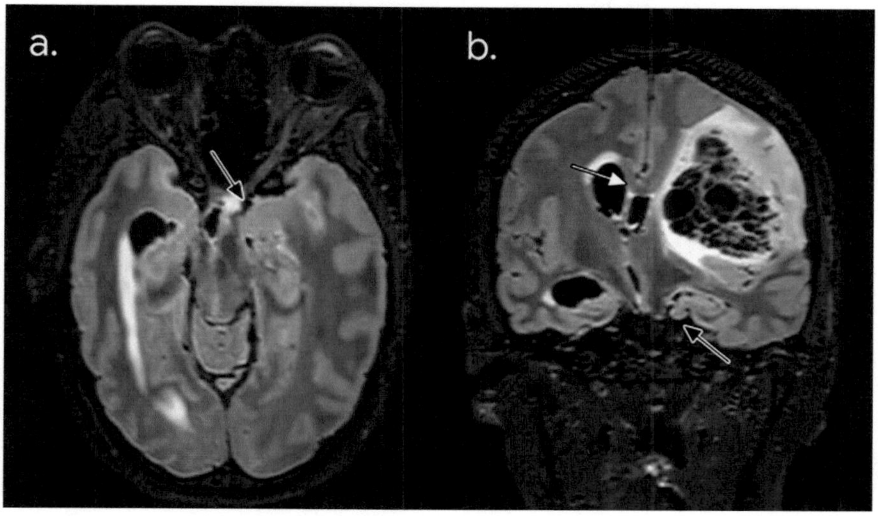

Figure 2.
FLAIR sequence of an MRI of the brain demonstrating uncal herniation (black arrows) and subfalcine herniation (white arrow), shown in axial (a) and coronal (b) orientations. This patient was found to have a WHO II oligodendroglioma.

Herniation type	Anatomical structures involved	Syndrome	Clinical considerations
Uncal	Mesial temporal lobe onto the oculomotor nerve, descending motor tracts, midbrain	Ipsilateral third nerve palsy, coma, contralateral weakness, or quadriparesis	Mass effect in the lateral supratentorial regions
Subfalcine	Cingulate gyrus, anterior cerebral artery	Decreased level of consciousness, contralateral lower extremity weakness	Mass effect in the medial supratentorial regions may cause anterior cerebral artery stroke
Tonsillar	Cerebellar tonsils onto the lower medulla, through the foramen magnum	Cardiorespiratory compromise, dysphagia	Mass effect in the cerebellum
Upward	Upper cerebellum onto the midbrain	Coma may result from a posterior cerebral artery or superior cerebellar artery stroke	May result from aggressive ventriculostomy drainage in the setting of a posterior fossa lesion
Transcalvarial	Dependent on location	Dependent on location	Herniation through a craniectomy site

Table 1.
Herniation syndromes and their anatomical and clinical characteristics.

indicates impending herniation is imminent and emergent action is needed. Importantly, herniation events can be precipitated by lumbar punctures or other CSF diversions in the presence of posterior fossa masses due to the generation of pressure changes. Because of this, imaging prior to lumbar punctures is recommended to further assess for this possibility, and if CSF drainage is still needed, 1 g/kg of 20% mannitol should be given immediately prior [7].

2.4 Management of acute intracranial hypertension

Treatment of intracranial hypertension aims to restore some aspect of homeostasis to the Monro-Kellie doctrine principles and centers on acutely reducing cerebral edema or restoring CSF flow. With regard to dangerously elevated ICP, independent of etiology, the Emergency Neurological Life Support (ENLS) protocol recommends the following (**Figure 3**) [11]:

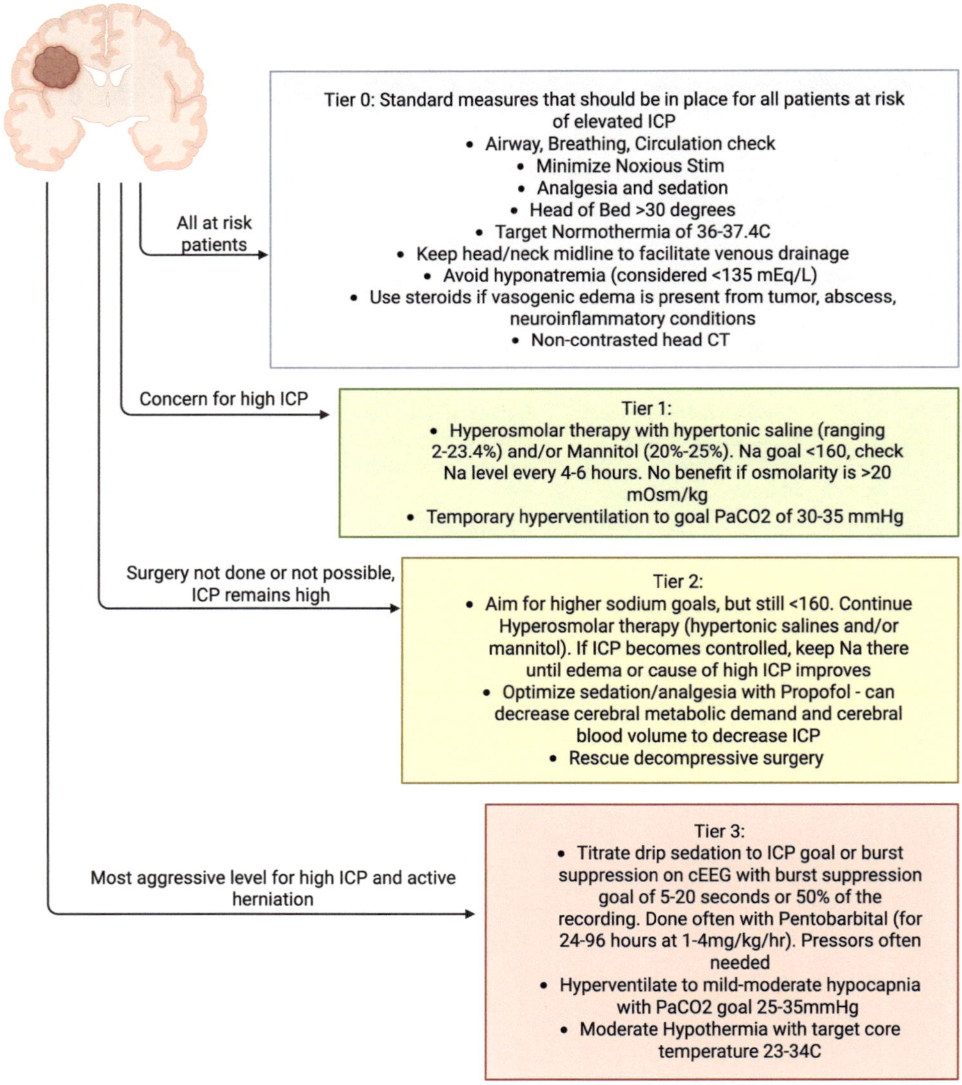

Figure 3.
An adapted tiered approach to elevated intracranial pressure management, as guided by the Emergency Neurologic Life Support (ENLS) protocols.

- Elevating the head of the bed to a minimum of 30–45 degrees to facilitate venous drainage.
- Hyperventilation with a goal PaCO2 of 25–30 mmHg to stimulate vasoconstriction via hypocapnia and lessen ICP [16].
- Pain control, including sedation and barbiturate coma if needed.
- Hyperosmolar therapies aim to create osmotic gradients intended to reduce interstitial fluid and sodium accumulation.
- Therapeutic hypothermia to decrease cerebral metabolic demand [17].
- Surgical interventions, if refractory to the above, which is often the case in malignancy-induced intracranial hypertension.

Hyperosmolar therapy is the cornerstone of cerebral edema and, thus, intracranial hypertension treatment and is comprised of osmotic agents such as 20% or 25% mannitol or hypertonic saline solutions concentrated as 3%, 7%, or 23.4%. Osmolar therapies use the impermeability of the blood–brain barrier to generate a gradient between the interstitial space and the intravascular space to mobilize water out of brain tissue and lessen cerebral edema. The reflection coefficient of a molecule is defined as its ability to cross a membrane, or, in the case of cerebral physiology, the blood–brain barrier. The reflection coefficient determines its effectiveness at generating this osmotic gradient, where 0 means full ability to cross and 1 means completely restricted. Mannitol has a reflection coefficient of 0.9, while hypertonic saline has a reflection coefficient of 1.0, making both very effective osmotic agents, with hypertonic saline slightly more robust. This gradient generation is effective in both cytotoxic and vasogenic edemas; however, it is reliant on an intact blood–brain barrier, which is often not present in intracranial malignancies due to their impacts on tight junctions, fenestration generation, and enhanced pilocytic astrocyte activity at the molecular level. Because of this, while they will still generate an osmotic gradient between brain tissue and the intravascular space, it will degrade over time as osmotic agents leak through the blood–brain barrier and into brain parenchyma. This is especially true with mannitol, which can subsequently lead to a rebound cerebral edema [7]. With regards to choosing which hyperosmolar therapy to use, due to side effect profiles, hypertonic saline is often preferred as it has less risk of acute kidney injuries and pulmonary edema compared to mannitol. Based on studies evaluating the effectiveness of 3% hypertonic saline versus 20% mannitol for management of elevated ICP, both modalities can lower ICP, with mannitol slightly more effective in the immediate period; however, hypertonic saline has a more sustained effect with a longer duration of ICP reduction [18].

With the vasogenic edema present around brain tumors, corticosteroids are also utilized to reduce edema and ICP treatment and may be more beneficial overall than osmolar therapy. Corticosteroids alleviate some of the inflammation in tumors that contributes to vascular permeability and help to support the integrity of the blood–brain barrier [7]. In both mouse and human models, steroids have demonstrated the ability to restore and preserve occludin function, thus maintaining tight junction stability and decreasing permeability [19]. In addition to helping reduce edema, corticosteroids help manage several additional aspects of cancer-related complications, including pain relief, reduction in tumor growth, appetite stimulation, and antinausea effects [20]. The favored steroid is dexamethasone due to its longer half-life, lower mineralocorticoid activity, high bioavailability, and lower risk of psychosis [7]. Two double-blind clinical trials examining the use of 4 mg per day and 8 mg per day of dexamethasone versus 16 mg per day in patients with brain metastasis concluded equivalent outcomes in the lower steroid doses, with high-dose-dependent toxic effects, suggesting lower doses to be safer overall unless there is concern for acute life-threatening elevated ICP with the potential for herniation, in which case 16 mg per day or more is warranted [20, 21]. One consideration is the impact steroids can have if CNS lymphoma has not been ruled out, as corticosteroids can lead to B lymphocyte lysis and confound biopsy results [9].

Refractory cerebral edema not responsive to hyperosmolar agents or steroids is not uncommon in intracranial malignancies and can be present in up to 39% of patients with vasogenic edema in the setting of tumors. Due to the influence of VEGF on blood–brain barrier permeability and the generation of edema, monoclonal antibodies targeting this molecule have been investigated for the management of refractory brain edema [22]. The most common medication for this is bevacizumab, which is a monoclonal antibody against VEGF-A, blocking the binding of VEGF-A to its receptor and impacting its effects on blood–brain barrier permeability and angiogenesis abilities (**Figure 4**). Recommended dosing is 5–10 mg/kg to achieve essentially unmeasurable levels of VEGF in serum. In one retrospective study evaluating refractory brain edema in metastatic lesions from either lung or colon cancer, there were statistically significant reductions in cerebral edema volume in 84.7% (61/72 patients) of patients and statistically and clinically significant improvement in symptoms in 90.3% (65/72 patients) of patients after only one 5 mg/kg dose of bevacizumab [22]. While adverse reactions do occur – including

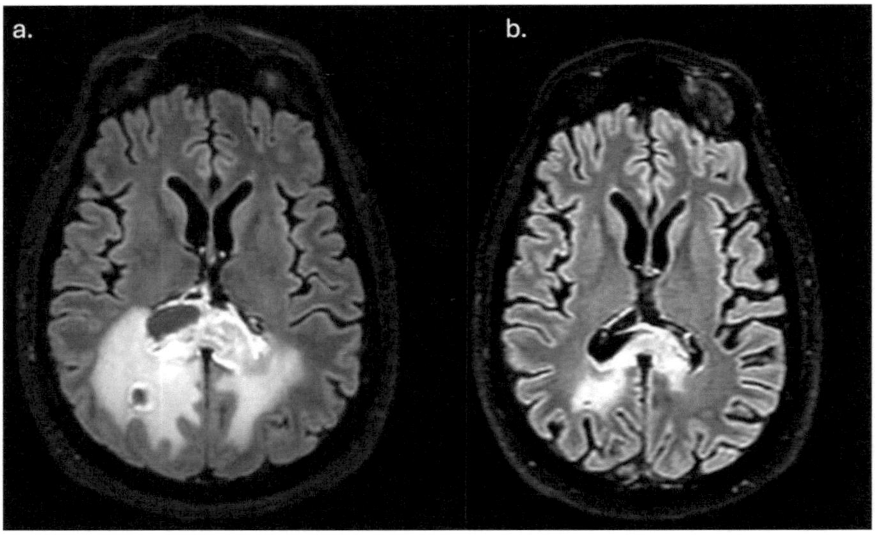

Figure 4.
Effects of bevacizumab on cerebral edema in neuro-oncologic tumors. A patient with known glioblastoma with significant bihemispheric cerebral edema (a), who then received one dose of bevacizumab 10 mg/kg, resulting in significant improvement in both tumor size and cerebral edema in one month (b).

hypertension, thrombocytopenia, thromboses (especially venous), sepsis, and intestinal perforation – none of the patients in this study had hemorrhages. Nine patients experienced hypertension, but this was associated with a significant tumor edema volume reduction, suggesting a potential correlation, and two patients had decreased platelet counts.

The mechanism of action of bevacizumab toward VEGF provides its effectiveness toward treating refractory cerebral edema and can be beneficial in the treatment of glioblastomas due to their abnormal angiogenesis pathophysiology from VEGF overexpression [23]. However, the side effects of bevacizumab in the immediate and long-term also stem from its target on VEGF, which is required in almost all tissues for normal development. In a 2014 study from the National Cancer Institute Malignant Glioma Trial Cohort analyzing 210 patients who received bevacizumab, 20% of patients were noted to have no side effects from the drug, while 11.9% experienced such severe toxicities that bevacizumab administration had to be stopped [23]. This included 1.9% of patients who died secondary to toxicity-related complications, such as stroke, neutropenia-related pneumonia, sudden unexplained deaths, and seizures [23].

With the role of VEGF in angiogenesis, hemorrhage and thrombosis are both possible complications associated with bevacizumab use. In the 2014 cohort, 28.1% of patients experienced hemorrhage, including intraparenchymal hemorrhage (IPH), with heightened risk increasing with the duration of medication use, although 19% of these cases were categorized as "minor bleeding" (epistaxis, GI hemorrhage, menorrhagia) [23]. While IPH can occur, in the National Cancer Institute cohort, these cases were asymptomatic, and prior studies have demonstrated that the hemorrhagic incidence with bevacizumab is either approximately equal to or less than that of patients who did not receive this treatment [24, 25].

Additional side effects reported with bevacizumab use have included hypertension, which increases almost exponentially with the longer duration of use, pancytopenia, colonic perforations, renal toxicities including proteinuria, and mild constitutional symptoms [23]. Although many side effects can occur, overall bevacizumab has been shown to be relatively well-tolerated, although expense remains a burden [26–28].

When cerebral edema reaches the threshold of impending herniation, surgical decompression is indicated to alleviate pressure before irreversible ischemic tissue damage can occur [7, 10]. This is often needed in malignancy-related causes of elevated ICP, as the underlying etiology of edema and herniation has not been controlled. Once patients are stabilized, surgical mass resection can be pursued for definitive management. The goal is gross total resection, although this is not always possible when tumors are significantly infiltrative, such as in GBMs [7, 11]. Based on a study by Stupp et al., for GBMs, debulking surgery followed by concomitant and adjuvant chemotherapy and radiation has been shown to meaningfully impact survival and quality of life without worsening toxicity [29]. Not all tumors will be amenable to surgical resection, including those located in the brainstem, basal ganglia, and thalamus; diffusely invasive tumors; those involving eloquent brain tissue; and multifocal lesions [10]. Importantly, radiation should not be instituted for patients with elevated ICP, as it can worsen intracranial hypertension from progressive edema [30]. When tumor location or mass effect leads solely to hydrocephalus without concern for additional sequelae, CSF diversion with either endoscopic third ventriculostomy or VPS is favored to restore flow dynamics [11].

3. Cerebrovascular complications

3.1 Arterial and venous thromboses

During the first month after a cancer diagnosis, the risk of stroke or TIA is increased upwards of six times compared to individuals without cancer, and at autopsy, 15% of cancer patients have evidence of cerebrovascular accidents (CVAs) [10]. Ischemic infarct risk is higher in certain types of metastatic cancers, including lung, pancreatic, and colorectal cancers, while hemorrhagic strokes are more common in hematologic malignancies, GBMs, oligodendrogliomas, and metastatic cancers known to bleed, including melanoma, lung, and breast cancers [7, 10, 23]. Lung adenocarcinoma has the highest association with stroke, followed by primary CNS malignancies, including gliomas, meningiomas, and primary CNS lymphoma [31]. While risk factors remain the same for strokes in cancer patients, many lack a conventional stroke mechanism (40%), and 18% have cryptogenic strokes [10]. Overall, the prognosis in patients with stroke and cancer is worse than in those without cancer, and they tend to be younger, with more severe strokes and higher rates of in-hospital mortality from separate ischemic events like myocardial infarctions [31].

There are several mechanisms at play with intracranial and systemic malignancies that lead to heightened cerebrovascular incidence and risk. Tumors can cause direct external compression of blood vessels through vessel invasion or vasogenic edema. In intravascular diffuse large B-cell lymphoma, tumors can infiltrate into the blood vessel walls, leading to vascular irregularities, which can subsequently cause occlusions or embolism from turbulent flow. Additionally, there can be vascular occlusion from tumor cells themselves, which can be seen in myxomas or lung cancers, from the impact of treatments such as radiation-induced atherosclerosis [32], as well as nonbacterial thrombotic endocarditis from the deposition of acellular aggregations of fibrin and platelets on cardiac valves, which can then embolize [10]. Nonbacterial thrombotic endocarditis is most often seen in adenocarcinoma patients. Systemically, cancers can lead to hypercoagulable or hyperviscous states. The hypercoagulable state of cancers comes from an increase in mucin production, which subsequently causes the formation of platelet-rich thrombi. In addition, tissue factors and procoagulant cytokines, including TNF-alpha, IL-1, and IL-6, are released, which potentiate clotting cascades [10]. Hematologic malignancies such as multiple myeloma, Waldenstrom's macroglobulinemia, chronic lymphocytic leukemia (CLL), and myelogenous leukemia result in hyperviscosity through increased stickiness of blood, thus increasing cerebral blood flow resistance and occlusion of small blood vessels. When hyperviscosity is present, patients will usually present with a triad of neurologic symptoms, visual disturbances, and mucosal bleeding [31].

The treatment of ischemic strokes in cancer patients is similar to that in noncancer patients, with a few notable exceptions. Thrombolytics such as tissue plasminogen activator/alteplase (tPA) and tenecteplase (TNK) are safe and efficacious for use in acute stroke patients with systemic cancers. However, both have relative contraindications for administration in cases of intracranial tumors due to increased bleeding risk, as well as in thrombocytopenia, which often accompanies cancers or cancer treatments [31]. Longterm, the use of antiplatelet and anticoagulation therapy is less clear. The increased venous thromboembolic (VTE) risk in cancer patients has been well established, but there is concern about an under-recognized link with arterial thromboembolism (ATE) as well. In VTEs, randomized clinical trials (RCTs) have shown noninferiority between direct oral anticoagulants (DOACs) and low molecular weight heparin (LMWH) for the prevention of venous clots up to 6-month follow-ups [33]. However, not much is known about the efficacy of anticoagulation in malignancy-related arterial clots, despite the known hypercoagulable state of cancers. A systematic review and meta-analysis of 14 RCTs led to the inclusion of 10,248 cancer patients with cancer and ATE events. Together, the difference between the pooled risk of arterial clots with anticoagulation use versus standard treatment or placebo was minimal (0.98% with anticoagulation; 1.48% with placebo or standard treatment, $p = 0.084$ and not statistically significant). Importantly, anticoagulation use was associated with a 1.6 times higher risk of major bleeding events and a 2.3 times higher risk of minor bleeding events [34].

The increased venous thromboembolism risk is not just present systemically in cancer patients; cerebral venous sinus thrombosis (CVST) risk is increased as well, with 1 in 10 patients with a CVST having a malignancy – a frequency that increases with age above 55 to one in four [7]. Like ischemic strokes from arterial clots, CVSTs more frequently occur within the first year following a cancer diagnosis, with higher incidence seen in hematologic, lung, breast, and gastrointestinal malignancies [7]. CVSTs present with symptoms consistent with high ICP, as well as seizures, focal neurologic deficits, and hemorrhagic or ischemic strokes (**Figure 5**) [11]. The etiologies typically include direct compression of the venous sinuses and the generation of a hypercoagulable state [31]. Often, CVST patients will require initiation of a heparin drip with monitoring for intracranial hemorrhage (ICH) development due to the availability of reversal agents and its shorter half-life, with subsequent transition to either DOACs or LMWH [7, 32].

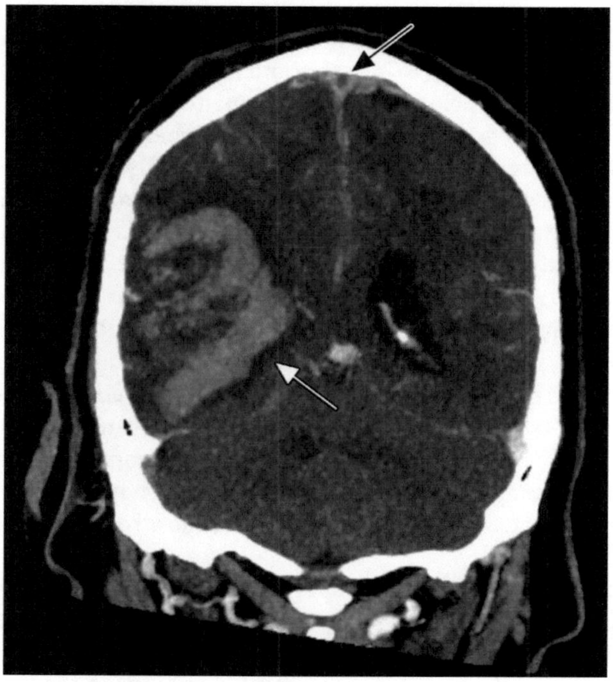

Figure 5.
A CT venogram demonstrating a large venous infarct with hemorrhage (white arrow) secondary to a venous sinus thrombosis within the sagittal sinus (black arrow).

3.2 Hemorrhagic complications

Unlike the array of direct and indirect mechanisms leading to ischemic cerebral arterial and venous events, the causes of malignancy-related ICHs are often either the tumor itself or a side effect of treatment. The predominant means of ICHin brain tumors are either intratumoral hemorrhage, caused by fragile neovascularization, tumor-infiltrated vasculature, and necrosis from outgrowing their blood supply; coagulopathy; or a combination of the two [35, 9]. The spontaneous ICHs seen in metastatic or primary brain cancers can be either intra or extraaxial. Overall, roughly anywhere from 1–10% of spontaneous IPHs are secondary to cancer, with IPHs occurring in both primary and metastatic tumors [7, 28, 35]. Most occur in patients with a known cancer and typically happen later in the disease course; however, 4% are the first manifestation of a malignancy [9]. Both primary CNS and metastatic cancers can bleed, with glioblastomas and oligodendrogliomas being the most common primaries, and melanoma, lung, breast, and renal cell carcinomas the most common metastatic cancers to hemorrhage. This is likely due to their grading with high cell turnover, making the tissue more fragile and friable [7].

Aside from the type of cancer and tumor size, specific treatments can also predispose patients to hemorrhage, such as biologics, chemotherapies, and radiation. Patients who are on Bevacizumab, the monoclonal antibody targeted to VEGF that is used to treat refractory vasogenic edema, have a three-fold increased risk of developing IPH despite its prevention of angiogenesis via VEGF inhibition [7, 35], although with a dose-dependent relationship. Chemotherapy-induced bone marrow suppression can lead to coagulopathies and pancytopenia, with thrombocytopenia increasing the risk of hemorrhage overall, including IPH [36]. This can also be seen in hematologic malignancies that similarly lead to impacted bone marrow production. Whole-brain radiation, utilized in instances of multiple intracranial lesions, generates friable blood vessels predisposed to bleeding and can result in the delayed development of malformed vasculature, including cavernous malformations and aneurysms [37].

Intratumoral hemorrhages present similarly to other intracranial bleeding, with headaches and hemiparesis being the most common (50%), followed by encephalopathy that can progress to coma, nausea and vomiting, and seizures [9]. Mainly intraparenchymal, hemorrhages in cancer patients can involve multiple hemispheres, especially when due to coagulopathies associated with cancer treatments or hematologic malignancies. Subarachnoid hemorrhages, subdural hemorrhages, and intraventricular hemorrhages without IPH can also develop [36]. Workup is similar to acute changes in elevated ICP, with a noncontrasted head CT to evaluate hemorrhage size and involved structures, followed by a contrasted MRI brain when stable. Vessel imaging with angiograms and venograms (CT and/or MR) can also be indicated when IPH is found and there is a concern for a tumor, but the MRI is unclear [38]. Imaging is needed to rule out alternative causes for IPH, including hemorrhagic transformation from ischemic stroke, venous thrombus, or cerebral amyloid angiopathy.

When tumors are the confirmed etiology of a hemorrhage, management is two-tiered. Initially, corticosteroids are utilized to alleviate any concurrent vasogenic edema. Once able, surgical removal of the tumor is indicated [38]. Per results from the International Surgical Trial in Intracerebral Hemorrhage (STICH), lobar hemorrhages – those located in the superficial cerebral gray or white matter – tend to have better surgical outcomes overall when compared to hematomas located >1 cm below the brain surface [39]. Intracranial and intratumoral hemorrhages can form clots, which can also be surgically evacuated if they are located superficially and are the cause of progressive symptoms; however, both the STICH and STICH II trials did not show improvement in patient outcomes from preemptive early evacuation of clots without symptomatic progression when compared to standard medical management [31, 9, 40]. Overall, short-term outcomes in cancer patients with intratumoral hemorrhages are in line with other etiologies of IPH; however, long-term outcomes are worse due to impacts from the underlying malignancy [36]. Mortality is high in these patients, with an estimated 31% mortality at one month, increasing to a 78% mortality rate at one year postintratumoral hemorrhage, potentially due to IPH occurring later in the cancer course [36, 41].

3.3 Pituitary apoplexy

A rare but life-threatening consequence of either ischemia or hemorrhage can be seen with pituitary apoplexy. Occurring in 2–12% of patients with pituitary adenomas, usually macroadenomas, in cancer it is usually the result of acute interruptions in blood supply to the tumor via a combination of tumor growth, reduction of blood flow to the tumor, and abnormal neovasculature [11]. While there are several other etiologies outside of neoplasia that can result in pituitary apoplexy, they are outside the scope of this chapter. Typically, the anterior pituitary (supplied by the superior hypophyseal artery) is primarily impacted without much involvement of the posterior pituitary (supplied by the inferior hypophyseal artery). Due to the location of the pituitary gland, pituitary apoplexy presents most commonly with acute severe retro-orbital headaches secondary to dural stretching and irritation of the meninges, nausea, cranial nerve palsies, which are seen in 52% of patients, and endocrinopathies [11, 9]. The oculomotor nerve is primarily involved, leading to visual abnormalities, ptosis, and mydriasis, while a bitemporal hemianopsia can develop if the mass extends upward to the optic chiasm. With cavernous sinus involvement, additional cranial nerves can become impacted, including the trochlear nerve, the abducens nerve, and the first and second divisions of the trigeminal nerve.

The endocrinopathies, especially acute adrenal insufficiency, are the principal causes of morbidity and mortality in patients with pituitary apoplexy. At least one anterior pituitary hormone will be dysfunctional in 80% of patients with pituitary apoplexy, the most common being adrenocorticotropic hormone (ACTH) at 70% [9]. ACTH helps regulate cortisol production, and its acute absence can lead to sudden hypotension and hyponatremia. Although possible, as mentioned previously, the posterior pituitary is less likely to be affected as its blood supply – the inferior hypophyseal artery – has some extracranial components that may not be impacted by tumor-related infarct or hemorrhage. For this reason, only 5% of patients with pituitary apoplexy will develop central diabetes insipidus [9].

4. Seizures

A common structural cause of a first-time seizure in the adult brain is a tumor [11]. Malignancies predispose individuals to seizures for a multitude of reasons, including cortical irritation, the presence of intratumoral hemorrhage, tumor characteristics, treatment side effects, and paraneoplastic conditions [7]. There are several molecular factors that are postulated to play a role in the generation of seizures in intracranial malignancies, including the isocitrate dehydrogenase (IDH)1 or IDH2 variation status. IDH is a driver mutation in low-grade gliomas, with IDH1 expressed in 70-80% of low-grade gliomas and only 5-10% of high-grade gliomas [10]. When mutated, IDH1 catalyzes isocitrate to 2-hydroxyglutarate (2-HG), a molecule that is structurally similar to glutamate, which is the excitatory neurotransmitter for NMDA receptors promoting epilepsy [7].

Unlike elevated ICP, where slower-growing tumors are more easily accommodated, slow-growing tumors are more likely to lead to new-onset seizures [10]. Roughly one-third of patients with primary or secondary brain tumors will develop tumor-associated epilepsy, with the most common tumors being: dysembryoplastic neuroepithelial tumors (80–100% seizure incidence); gangliomas (80–100% seizure incidence); diffuse low-grade gliomas such as oligodendrogliomas and oligoastrocytomas (60–80% seizure incidence); GBMs (40–60% seizure incidence); and, lastly, brain metastases (20–40% seizure incidence), with melanoma being the most likely cause of metastatic seizure focus [10].

Malignancy-associated epilepsy tends to occur early in the disease course, with the risk of occurrence inversely proportional to the WHO grade of the tumor, while tumor-related status epilepticus is directly related to the WHO grade, seen more frequently in more malignant cancers and metastases [10]. The exact semiologies witnessed vary depending on tumor location and the structures impacted; however, the general classifications seen include focal seizures with preserved awareness (23–58%), focal impaired seizures (7–31%), focal seizures with secondary generalization to spread to the entire brain (10–68%), and status epilepticus, which is defined as a seizure lasting longer than five minutes or multiple seizures in short duration without a return to baseline in between [10]. Roughly 15–22% of the patients who initially develop tumor-related epilepsy will progress to status epilepticus, typically a manifestation later in the disease course, and approximately 3–12% of all status epilepticus cases stem from brain tumors, although tumors lend more to focal status epilepticus [42]. This is lower than the general population of noncancer patients with epilepsy, who have a reported incidence of status epilepticus approaching 30–40%, but may also be underreported due to their occurrence toward the end of the disease (at times of palliative or hospice care) and with the potential for unrecognized nonconvulsive status epilepticus (NCSE) [10, 31]. There is a relatively even split between patients being in status epilepticus at the time of their tumor diagnosis (29%), during progression of their disease (23%), and during perceived times of disease stability (23%). While status epilepticus itself requires clinical findings, NCSE can be a more subtle finding and may be the main cause of encephalopathy in up to 6% of patients with cancer [10].

Greater than 10% of neurologic causes of death in cancer patients stem from refractory seizures, although interestingly, there have been studies suggesting longer survival in GBM patients when their diagnosis is made following seizure presentation [31]. In intracranial malignancies, either primary or metastatic, seizures can be difficult to control through medications alone and often remain refractory without surgical resection due to their functioning as a structural seizure focus. With surgical resection, seizure control can be accomplished in 68–85% of cases [11], especially when gross total resection is possible [43]. In one study by Chang et al. in 2008, 81% of their patients with low-grade gliomas presented with seizures, likely due to the association between oligodendrogliomas and cortical locations. Of these patients, 50% had uncontrolled seizures in the preoperative setting despite appropriate medications, with worse control associated with focal seizures, temporal locations, and longer preoperative seizure history. Following surgery, 98% of patients were able to remain seizure-free at six months if they had gross total resection of their mass, as well as a preoperative seizure history of less than one year and a more complex seizure semiology. If none of these factors were present, only 44% of patients had postoperative seizure freedom [43].

However, seizure risk is also heightened in the immediate postoperative setting, extending up to 48 hours after surgery, in 1–12% of patients [7]. While resection is largely curative from the seizure standpoint, antiseizure medications (ASMs) remain a cornerstone of medical management in malignancy-related epilepsy and are often needed prior to resection or in combination with surgical options to maximize seizure control. The ASMs available for use in cancer patients are largely similar to those used in noncancer patients; however, medication interactions with chemotherapeutics need to be considered. The first-line agent, due to its safety profile, tolerability, efficacy, and lack of medication interactions, is levetiracetam [10]. In one study of 176 patients, 91% achieved seizure freedom with the use of levetiracetam, without lab abnormalities [44]. This agent alone may not be sufficient, and additional medications may be required. Newer formulations, such as lamotrigine or lacosamide, are second-line options as they lack interactions with chemotherapeutics. Valproic acid may be used, especially in cases of status epilepticus when intravenous loading of medications is required, but it is important to note that it is a cytochrome P450 (CYP450) inhibitor. Many chemotherapies are metabolized via the CYP450 pathway and thus can be elevated to toxic levels with the addition of valproic acid [10]. Conversely, phenytoin, phenobarbital, and carbamazepine are avoided in these patient populations as they are CYP450 inducers and can lead to increased metabolism of chemotherapies.

First-line ASMs, with or without resection, may not be enough to control seizures in tumor patients. Approximately 60% of first-line ASMs fail in tumor-associated epilepsy management, with resistance seemingly associated with higher tumor grade and tumor progression [45]. One hypothesis centers on exo-transportation of ASMs from the area of the brain in which the tumor is located, due to upregulation and overexpression of multidrug resistance (MDR) proteins like glycoprotein-P on tumors, thus leading to subtherapeutic parenchymal levels of ASMs [45]. Interestingly, status epilepticus stemming from malignancies appears to be more responsive to ASMs, with only 14–18% refractory compared to tumor-related epilepsy [45]. While more of a rarity for malignancy-related status epilepticus, pharmacoresistance does occur, meaning two or more appropriately chosen and dosed antiseizure medications are not enough to stop the seizures. This is also termed "super refractory status epilepticus," and its severity is directly related to its duration [46]. When pharmacoresistant seizures or status epilepticus is present, the first-line approach is adjusted and pivots to rapidly uptitrated anesthetics such as midazolam, pentobarbital, propofol, or ketamine infusions for 24–48 hours [46]. Antiseizure medications are often added at this time as well for more durable long-term control. After 48 hours, CNS sedatives are weaned off; however, ongoing seizures can occur, at which time anesthetics are resumed, third-line anticonvulsants are escalated, and the following are considered: immunotherapies (steroids, intravenous immunoglobulins); electric and magnetic stimulation therapies; ketogenic diets; magnesium or pyridoxine infusions; CSF drainage; or surgical resection [46]. When this point is reached, prognosis is poor, with long-term mortality approaching 30–50%; however, this is lower when super refractory status epilepticus arises from a tumor focus [46].

5. Spinal cord compression

With the high vascularity of the spinal cord, most tumors found here are metastatic, spreading from primary cancers such as prostate, breast, kidney, gastrointestinal tract, thyroid, non-Hodgkin's lymphoma, and multiple myeloma [7]. They predominate in the thoracic cord (70%), followed by the lumbar (20%) and cervical (10%) [10]. Upwards of 60% of the tumors in the spinal cord, especially the metastatic lesions, are extradural, while 30% are located intradural extramedullary – namely meningiomas, nerve sheath tumors, and drop metastases – and 10% are intradural intramedullary, such as ependymomas and astrocytomas found in the cervical cord regions [7]. Most often, cancerous cells from a systemic primary tumor travel to the spinal epidural space through hematogenous spread to the vertebral bodies, given their high vascularity (85% of the time), although there is a small percentage (15%) that enters through the intervertebral foramen, as seen in lymphomas and neuroblastomas [10]. As tumor cells grow within vertebrae, they destroy the bone through expansion, outgrowing into the epidural space and causing compression of the spin al cord.

Seen in approximately 5–10% of patients with cancer, the symptoms of metastatic spinal cord compression include sharp, shooting, and burning back pain that starts deep in the vertebral bodies; radiculopathy from nerve root compression; motor and sensory deficits, including quadri- and paraplegias; and autonomic instabilities, such as orthostatic hypotension and bowel/bladder dysfunction [10]. Once clinically identified, imaging with a contrasted MRI is the gold standard, showing a 95% accuracy in detecting spinal malignancies when the whole spine is imaged. An acceptable alternative, if needed, is CT myelography; however, plain spinal CTs and radiographs are not sufficient. On MRI, masses in the vertebrae will appear hypointense on T1, hyperintense on T2, and contrast-enhancing due to their vascularity [47], while the spine itself will show evidence of venous congestion or ischemia [10]. Short Tau Inversion Recovery (STIR) sequences on MRI can also be beneficial when bony metastases are predominant, as they suppress the fat of the bone marrow and allow for more accurate visualization of abnormal tissues [47].

Due to the risk of rapid deterioration and the high stakes of potentially significant neurologic dysfunction, spinal cord compression from malignancy is a medical and surgical emergency that, once identified, must be addressed immediately. The number of patients with cancer experiencing metastatic spinal cord disease is increasing as advancements in cancer care allow for prolonged survival, and fast recognition and action are paramount. Emergent treatment can depend on a variety of factors, including overall patient prognosis, baseline status prior to spinal cord compression, radiographic assessment per the Neurologic, Oncologic, Mechanical, and Systemic (NOMS) framework, and assessment of spinal instability from neoplasm [47]. The epidural spinal cord compression (ESCC) scale is a radiologic grading system based on the presence or absence of thecal sac or spinal cord compression/deformation from malignancy, with grades 0, 1a, and 1b considered appropriate for radiotherapy, while grades 2–3 should undergo surgical decompression prior to radiation [47]. Similarly, and often used in conjunction with the ESCC scale, the Spinal Instability Neoplastic Score (SINS) is an 18-point scoring system utilizing imaging and clinical symptoms, such as pain, to determine spinal stability. A score of 7 or higher (potentially unstable or unstable) indicates surgical consultation for potential intervention [47].

Treatment of malignancy-related spinal cord compression can be categorized as palliation, medication, surgery, or radiation. These can be combined or follow one another to maximize neurologic outcomes. A cornerstone of medical management in malignant ESCC is steroids, used similarly in the management of spinal cord compression as for vasogenic edema, with the goal of reducing swelling. When hematologic primaries present, they can have an oncolytic feature that can additionally assist with reducing cord compression [48]. Recommended dosing is high, around 4 mg three times daily or four times daily, for a maximum dosing of around 16 mg per day, and higher doses can demonstrate toxicity without improving ambulatory status [48]. When given in conjunction with radiotherapy, steroids can improve outcomes compared to radiotherapy alone [49].

Radiotherapies have been used for some time as a treatment for malignant ESCC to help with the improvement of neurologic function, as well as the reduction of pain control from a palliative aspect and a quality-of-life improvement standpoint. Radiotherapies can include either conventional external beam radiotherapy (cERBT) or stereotactic ablative radiotherapy (SABR). Current available data suggest that the duration of treatment should depend on prognosis; shorter courses should be considered in cases with poor prognosis, and longer durations in instances of more favorable outcomes due to better local control. Radiotherapy can be utilized in conjunction with surgery, often with surgery done prior to radiation with the goal of debulking the area with operative decompression, then radiating for local tumor control [48, 50].

While previous reports have suggested poor-to-no neurologic improvement in patients with no motor functioning for 48 hours or more [51], despite an average time to intervention of approximately 14 days [52], a more recent, albeit small, retrospective study of 37 patients with malignant ESCC showed that 59.5% of patients regained the ability to ambulate after surgical decompression of the spinal cord, with a nine-day median of preoperative paralysis prior to intervention [52]. On meta-analyses, data continue to support that the earlier the intervention, the more likely the neurologic recovery [53], with larger studies even suggesting better American Spinal Injury Association (ASIA) score improvement if surgery is performed within 16 hours of admission [49]. Altogether, the recommendations continue to center on surgical decompression as soon as possible, although surgery should not be avoided just because a patient is outside of the standard 48-hour window.

6. Treatment-induced emergencies

Due to the intensity and aggressive nature of their mechanisms of action, treatments for neuro-oncologic diseases can lead to life-threatening complications, including cerebral edema, infections, posterior reversible encephalopathy syndrome (PRES), status epilepticus, cerebrovascular complications, and neurotoxicity [7]. All mechanisms of cancer management, from radiation to chemotherapies, biologics, and hematopoietic stem cell transplantations, can lead to these neurologic emergencies [7].

Recently, chimeric antigen receptor (CAR) T cells have become more widely recognized as an effective treatment option for cancers, especially advanced hematologic malignancies. In this method, DNA encodes an extracellular tumor antigen receptor that is then linked to an intracellular component designed to mediate T-cell activation, thus targeting T-cells to specific tumor molecules for destruction [54]. Due to their T-cell activation mechanism, there is a strong supraphysiologic immune response that can target healthy tissues as well. With this, the most common systemic adverse effects from CAR-T therapies are cytokine release syndrome (CRS), cytopenias, and neurotoxicity, which is seen in 70% of patients and can range from mild confusion to fatal edema [55].

There are two key neurotoxicological consequences of CAR-T therapies: immune effector cell-associated neurotoxicity syndrome (ICANS) and tumor inflammation-associated neurotoxicity (TIAN). There is another, more mild neurologic entity considered non-ICANS neurotoxicity (NINTs), which presents with cranial and peripheral neuropathies, cognitive complaints, and parkinsonism but is overall outside the scope of this chapter. ICANS can manifest in a variety of presentations, ranging from mild to moderate (grades 2–3) with acute behavioral changes, disorientation, tremors, and aphasia, to severe (grades 3–4) with life-threatening cerebral edema impacting the level of consciousness [56]. This severity is linked to neuroaxonal damage, as suggested by evidence of elevated serum neurofilament light chains (NfL) correlating with ICANS severity [57]. Regardless of grade, ICANS is typically an acute, monophasic condition occurring on average five days after CAR-T infusion, although it can happen up to three weeks later [58, 59].

Currently, neurologic toxicity in ICANS is hypothesized to stem from inflammation caused by cytokines released from proliferating T cells – IL-1, IL-6, IL-15, granulocyte-macrophage colony-stimulating factor – which is supported by the almost parallel presence of CRS with ICANS presentations [54]. Importantly, the cytokines IL-1 and IL-6 have implications for the blood–brain barrier, increasing its permeability and potentially allowing for the entry of CAR-T cells into the CNS, contributing to or causing the neurotoxicity that is observed [54]. However, while autopsies of ICANS patients have shown dilapidated blood–brain barriers, they often lack evidence of actual CAR-T cell infiltration into the brain tissue itself [58]. Elevated IL-1 has been associated more with neurologic symptoms compared to IL-6, which is seen elevated in CRS, suggesting that treatment of ICANS with anakinra, an IL-1 receptor antagonist, is more beneficial than tocilizumab, which impacts IL-6, as anakinra is able to cross the blood–brain barrier [56, 60]. Similarly, the earlier onset and longer duration of CRS in these patients increases the likelihood of ICANS development, supporting the need for CRS treatment when detected with medications such as tocilizumab [61]. Caution is recommended with tocilizumab use in ICANS as it should not be used in isolation – while helpful in the treatment of CRS due to its target molecule, it does not effectively cross the blood–brain barrier and thus may be ineffective in neurotoxicities [61]. Anticytokine medications should be used in conjunction with steroids, especially in the setting of cerebral edema. Interestingly, statins have been proposed to reduce ICANS incidence through stabilization of the endothelium and their antiinflammatory properties [62].

The designation, monitoring, and management of ICANS are based on a symptomatic scoring system called the Immune Effector Cell Encephalopathy Score, or ICE score, which should be administered at least twice daily for patients undergoing CAR-T cell therapy and continued out to 30 days from the time of the infusion. The ICE score is a composite of orientation scoring a total of four points, naming for a total of three points, following commands with a total of one point, writing for one point, and attention for one point; the higher the ICE score, the less severe the ICANS. When a patient has reached Grade 2 (ICE score 3–6), they require the initiation of corticosteroids, and by Grade 3 (ICE score 0–2), they require seizure prophylaxis with Keppra and routine imaging every two to three days to monitor for developing cerebral edema until symptoms resolve. By Grade 4 (ICE score 0, comatose), high-dose steroids with intravenous methylprednisolone are needed, as well as elevated ICP management with acetazolamide, elevation of the head of the bed, hyperventilation, and osmolar therapies due to the presence of life-threatening cerebral edema and potentially seizures (**Table 2**) [63, 64]. Regardless of ICE score or ICANS grade, thiamine supplementation is recommended, as deficiencies in this vitamin can be seen in inflammatory states (such as sepsis) and can decrease the stability of the blood–brain barrier [58].

ICANS grade	ICE score	Symptoms	Management
Grade 1	7–9	• Awake and nonfocal • Fatigued	Supportive care IV hydration EEG MRI
Grade 2	3–6	• Wakens to voice • Delirious • Somnolent	Supportive care Consider transfer to ICU Start antiepileptic medication – first line agent, usually Keppra. Start low-dose steroid – dexamethasone 10 mg.
Grade 3	0–2	• Requires tactile stimulation to wake • Seizures – focal or generalized – that resolve quickly. • Nonconvulsive status on EEG that resolves immediately with intervention. • Local edema on imaging can be present.	Supportive care ICU Continuous steroids - Dexamethasone 10 mg q6hrs Antiseizure medication – first-line agent is Keppra Consider repeating MRI every 2–3 days.
Grade 4	0	• Not rousable/Comatose • Status Epilepticus* • Motor weakness – hemiparesis or paraparesis is possible. • Diffuse cerebral edema on imaging	Supportive care ICU Hyperosmolar therapies – hypertonic saline, mannitol status epilepticus management – antiseizure medications (usually two antiseizure medications, e.g., Keppra and Lacosamide), anesthetic agents (e.g., propofol, midazolam, ketamine). Consider immunotherapies Consider repeating MRI every 2–3 days.

**Status epilepticus is defined as generalized seizures lasting longer than five minutes or multiple seizures without a return to baseline in-between.*

Table 2.
ICANS management recommendations differ depending on the grade, which is based on the ICE score and the symptoms experienced by the patient.

Global cerebral dysfunction is the prominent feature of ICANS and underlies its symptomatology, believed to be due to an on-tumor, off-target mechanism. In contrast, TIAN is a more regional phenomenon, with complications localized to the tumor site due to an on-tumor, on-target toxicity. TIAN can be observed in therapies other than CAR-T treatments, including immune-checkpoint inhibitors, antibody-based therapies, cytokine-based therapies, and cancer-related vaccines [56]. Similar to ICANS, there is a grading system based on deficit severity, ranging from mild (grade 1) to life-threatening (grade 4), to death (grade 5). TIAN may manifest as two separate clinical phenotypes. In TIAN type 1, inflammatory edema arising from the tumor site directly leads to mass effect and subsequent complications, including obstructive hydrocephalus and potential herniation [56]. TIAN type 2, however, results from an abnormality in local neuronal-immune interactions without the presence of edema, leading to a worsening of known neurologic symptoms and the potential for electrophysiologic disturbances [56]. The recognition of TIAN is important not only due to the potential for acute neurologic changes but also because it encompasses symptoms that were previously considered pseudoprogression or delayed radiation-related necrosis [56].

Neurotoxicity from radiation can occur in three stages, ranging from acute to early delayed, to long-term delayed. Acute radiation neurotoxicity, occurring within days of treatment, can progress to neurologic emergencies in the form of enhanced cerebral edema due to damage to capillary wall integrity. As a result, the leaky capillaries lead to extravasation of fluid and proteins, causing subsequent complications of edema [7]. Patients present with symptoms similar to the clinical manifestations of vasogenic edema, including headache, somnolence, nausea and vomiting, and worsening neurologic deficits. This condition is easily managed, with a good response to steroids and an overall favorable prognosis [10, 11]. It is also possible for delayed radiation neurotoxicity to present with cerebral edema, generated via relentless injury and destruction of oligodendrocytes and neuronal epithelium, creating an inflammatory cascade that can ultimately result in edema [7]. This condition is similarly responsive to management with corticosteroids, as is acute radiation necrosis. In both instances, as with the vasogenic cerebral edema associated with tumors themselves, treatment with the anti-VEGF monoclonal antibody bevacizumab can be utilized [11].

7. Conclusion(s)

Prompt recognition of neuro-oncologic emergencies is vital to ensuring a maximal quantity and quality of life for those with neoplastic disease involving the CNS. These patients carry a high risk of direct injury from the tumor, related to cerebral edema, hydrocephalus, and herniation, but also may be prone to thrombotic disease and seizures. With the advent and expanded use of CAR-T therapy, those who manage neurologically critically ill patients must also be familiar with ICANS and its management. Though patients with neurooncologic disease represent a vulnerable group, some aspects of their acute management must be extrapolated from those of other disease processes, as they represent a minority of instances of severe presentations, in contrast to more commonly encountered presentations of herniation syndromes, status epilepticus, and hydrocephalus.

Conflict of Interest

The authors declare no conflicts of interest.

Author details

Alicia Dupre and Andrew Treister*
Department of Neurology, Section of Neurocritical Care, Oregon Health & Science University, Portland, OR, USA

*Address all correspondence to: treistea@ohsu.edu

References

[1] Cancer Tomorrow [Internet]. Available from: https://gco.iarc.who.int/tomorrow/en/dataviz/isotype?types=1&single_unit=10000&cancers=31 [Accessed: 2026-February-12]

[2] Bray F, Laversanne M, Sung H, Ferlay J, Siegel RL, Soerjomataram I, et al. Global cancer statistics 2022: GLOBOCAN estimates of incidence and mortality worldwide for 36 cancers in 185 countries. CA: A Cancer Journal for Clinicians. 2024;**74**:229–263

[3] Bray F, Laversanne M, Weiderpass E, Soerjomataram I. The ever-increasing importance of cancer as a leading cause of premature death worldwide. Cancer. 2021;**127**(16):3029–3030

[4] SEER [Internet]. Cancer of the Brain and Other Nervous System - Cancer Stat Facts. Available from: https://seer.cancer.gov/statfacts/html/brain.html [Accessed: 2026-February-12]

[5] Habbous S, Forster K, Darling G, Jerzak K, Holloway CMB, Sahgal A, et al. Incidence and real-world burden of brain metastases from solid tumors and hematologic malignancies in Ontario: A population-based study. Neuro-Oncology Advances. 2020;**3**(1):vdaa178

[6] Tabouret E, Chinot O, Metellus P, Tallet A, Viens P, Goncalves A. Recent trends in epidemiology of brain metastases: An overview. Anticancer Research. 2012;**32**(11):4655–4662

[7] Maciel CB, Busl KM. Neuro-oncologic Emergencies. Continuum. 2024;**30**(3):845–877

[8] Benson JC, Madhavan AA, Cutsforth-Gregory JK, Johnson DR, Carr CM. The monro-kellie doctrine: A review and call for revision. American Journal of Neuroradiology. 2022;**44**(1):2–6

[9] Ehresman JS, Bettegowda C. Treatment of Oncologic Emergencies in the Neurocritical Care Unit. Nelson SE, Nyquist PA. Neurointensive Care Unit: Clinical Practice and Organization. Cham, Switzerland: Springer International Publishing; 2020. pp. 185–193. DOI: 10.1007/978-3-030-36548-6_14

[10] Jo JT, Smith D. Handbook of Clinical Neurology. Wijdicks E, Kramer A, editors. Vol. **141**. Amsterdam, Netherlands: Elsevier; 2017

[11] Suarez-Meade P, Marenco-Hillembrand L, Sherman WJ. Neuro-oncologic Emergencies. Current Oncology Reports. 2022;**24**(8):975–984

[12] Balestreri M, Czosnyka M, Hutchinson P, Steiner LA, Hiler M, Smielewski P, et al. Impact of intracranial pressure and cerebral perfusion pressure on severe disability and mortality after head injury. Neurocritical Care. 2006;**4**:8–13

[13] Mount CA, Das JM. Cerebral Perfusion Pressure. In: StatPearls. Treasure Island, FL, USA: StatPears Publishing. 2023

[14] Kaal ECA, Vecht CJ. The management of brain edema in brain tumors. Current Opinion in Oncology. 2004;**16**(6):593–600

[15] Mayhan WG. VEGF increases permeability of the blood-brain barrier

via a nitric oxide synthase/cGMP-dependent pathway. American Journal of Physiology-Cell Physiology. 1999;**276**(5):C1148–53

[16] Zhang Z, Guo Q, Wang E. Hyperventilation in neurological patients. Current Opinion in Anaesthesiology. 2019;**32**(5):568–573

[17] Flynn LMC, Rhodes J, Andrews PJD. Therapeutic hypothermia reduces intracranial pressure and partial brain oxygen tension in patients with severe traumatic brain injury: Preliminary data from the eurotherm3235 trial. Therapeutic Hypothermia and Temperature Management. 2015;**5**(3):143–151

[18] ThongrongC TW, Kasemsiri P, Duangthongphon P, Kitkhuandee A, Plailaharn N, et al. Comparison of utilizing a hypertonic saline solution and mannitol to improve brain relaxation during craniotomy in patients with brain tumours: A prospective randomized controlled trial. Scientific Reports. 2025;**15**(1):30912

[19] Witt KA, Sandoval KE. Advances in Pharmacology. Davis TP, editor. Vol. **71**. Cambridge, MA, USA: Academic Press; 2014

[20] Shahrabi M, Abdollahi A, Lahiji MN, Sorbi T, Solduzian M. Effects of steroids on vasogenic edema due to brain tumors, a narrative review. Interdisciplinary Neurosurgery. 2025;**41**:102106

[21] Ryken TC, McDermott M, Robinson PD, Ammirati M, Andrews DW, Asher AL, et al. The role of steroids in the management of brain metastases: A systematic review and evidence-based clinical practice guideline. Journal of Neuro-Oncology. 2009;**96**:103–114

[22] Bai X, Zhou M. The benefit of bevacizumab therapy in patients with refractory vasogenic edema caused by brain metastasis from lung and colon cancers. Frontiers in Oncology. 2022;**12**:838670

[23] Odia Y, Shih JH, Kreisl TN, Fine HA. Bevacizumab-related toxicities in the National Cancer Institute malignant glioma trial cohort. Journal of Neuro-Oncology. 2014;**120**(2):431–440

[24] Khasraw M, Holodny A, Goldlust SA, DeAngelis LM. Intracranial hemorrhage in patients with cancer treated with bevacizumab: The Memorial Sloan-Kettering experience. Annals of Oncology. 2012;**23**(2):458–463

[25] Besse B, Lasserre SF, Compton P, Huang J, Augustus S, Rohr UP. Bevacizumab safety in patients with central nervous system metastases. Clinical Cancer Research. 2010;**16**(1):269–278

[26] Dixit K, Singer L, Grimm SA, Lukas RV, Schwartz MA, Rademaker A, et al. A phase ii trial of bevacizumab in patients with recurrent/progressive solid tumor brain metastases that have progressed following whole-brain radiation therapy. Cancers. 2024;**16**:2133

[27] Francini E, Mazzaroppi S, Salvati M, Caponnetto S, Laera L, Migali C, et al. P17.32 * efficacy and tolerability of bevacizumab monotherapy in patients with previously treated recurrent glioblastoma multiforme (GBM). Neuro-Oncology. 2014;**16**:ii94–ii94

[28] Sharma A, Low J, Mrugala MM. Neuro-oncologists have spoken – The role of bevacizumab in the inpatient setting. A clinical and economic conundrum. Neuro-Oncology Practice. 2018;**6**(1):30–36

[29] Stupp R, Mason WP, van den Bent MJ, Weller M, Fisher B, Taphoorn MJB, et al. Radiotherapy plus concomitant and adjuvant temozolomide for glioblastoma. New England Journal of Medicine. 2005;**352**:987–996

[30] Pater K, Puskulluoćlu M, Zygulska AL. Oncological emergencies: Increased intracranial pressure in solid tumours' metastatic brain disease. Przegląd Lekarski. 2014;**71**(1):91–94

[31] Threlkeld ZD, Scott BJ. Neuro-Oncologic Emergencies. Neurologic Clinics. 2021;**39**(2):545–563

[32] Lyman GH, Carrier M, Ay C, et al. American Society of Hematology 2021 guidelines for management of venous thromboembolism: Prevention and treatment in patients with cancer. Blood Advances 2021;**5**:927–974

[33] Schrag D, Uno H, Rosovsky R, Rutherford C, Sanfilippo K, Villano JL, et al. Direct oral anticoagulants vs low-molecular-weight heparin and recurrent VTE in Patients with cancer: A randomized clinical trial. JAMA. 2023;**329**:1924–1933

[34] Xu Y, Cole K, Collins E, Moledina A, Mallity C, Carrier M. Anticoagulation for the prevention of arterial thrombosis in ambulatory cancer patients. JACC: CardioOncology. 2023;**5**(4):520–532

[35] Lim S, Clarke NH, Maloney SL, Sener UT, Caron SJ, Kizilbash SH, et al. Bevacizumab exerts dose-dependent risk for intracranial hemorrhage in patients with malignant gliomas. Journal of Neuro-oncology. 2025;**172**:273–280

[36] Velander AJ, DeAngelis LM, Navi BB. Intracranial Hemorrhage in Patients with Cancer. Current Atherosclerosis Reports. 2012;**14**(4):373–381

[37] Rogers LR. Cerebrovascular complications in cancer patients. Neurologic Clinics. 2003;**21**(1):167–192

[38] Kuzeyli K, Çakir E, Usul H, Karaarslan G, Yazar U, Baykal S, et al. Intratumoral haemorrhage: A clinical study. Journal of Clinical Neuroscience. 2004;**11**:490–492

[39] Mendelow A, Gregson B, Fernandes H, Murray G, Teasdale G, Hope D, et al. Early surgery versus initial conservative treatment in patients with spontaneous supratentorial intracerebral haematomas in the International Surgical Trial in Intracerebral Haemorrhage (STICH): A randomised trial. The Lancet. 2005;**365**:387–397

[40] Mendelow AD, Gregson BA, Rowan EN, Murray GD, Gholkar A, Mitchell PM. Early surgery versus initial conservative treatment in patients with spontaneous supratentorial lobar intracerebral haematomas (STICH II): A randomised trial. The Lancet. 2013;**382**(9890):397–408

[41] Navi BB, Reichman JS, Berlin D, Reiner AS, Panageas KS, Segal AZ, et al. Intracerebral and subarachnoid

hemorrhage in patients with cancer. Neurology. 2010;**74**:494–501

[42] Tziakouri A, Hottinger AF, Novy J, Rossetti AO. Status epilepticus management in patients with brain tumors. A cohort study. Seizure: European Journal of Epilepsy. 2024;**120**:1–4

[43] Chang EF, Potts MB, Keles GE, Lamborn KR, Chang SM, Barbaro NM, et al. Seizure characteristics and control following resection in 332 patients with low-grade gliomas. Journal of Neurosurgery. 2008;**108**:227–235

[44] Rosati A, Buttolo L, Stefini R, Todeschini A, Cenzato M, Padovani A. Efficacy and Safety of Levetiracetam in Patients With Glioma. Archives of Neurology. 2010;**67**(3):343–346

[45] Goonawardena J, Marshman LAG, Drummond KJ. Brain tumour-associated status epilepticus. Journal of Clinical Neuroscience. 2015;**22**(1):29–34

[46] Kirmani BF, Au K, Ayari L, John M, Shetty P, Delorenzo RJ. Super-Refractory Status Epilepticus: Prognosis and Recent Advances in Management. Aging and Disease. 2021;**12**(4):1097

[47] Bai J, Grant K, Hussien A, Kawakyu-O'Connor D. Imaging of metastatic epidural spinal cord compression. Frontiers in Radiology. 2022;**2**:962797

[48] Akanda ZZ, McKay MJ. Narrative review—diagnosing and managing malignant epidural spinal cord compression: An evidence-based approach. Annals of Translational Medicine. 2023;**11**(11):386–386

[49] Sørensen S, Helweg-Larsen S, Mouridsen H, Hansen HH. Effect of high-dose dexamethasone in carcinomatous metastatic spinal cord compression treated with radiotherapy: A randomised trial. European Journal of Cancer (Oxford, England: 1990). 1994;**30A**(1):22–27

[50] Bilsky MH, Angelov L, Rock J, Weaver J, Sheehan J, Rhines L, et al. Spinal radiosurgery: A neurosurgical perspective. Journal of Radiosurgery and SBRT 2011;**1**:47–54

[51] Robson P.Metastatic spinal cord compression: A rare but important complication of cancer. Clinical Medicine (London, England). 2014;**14**(5):542–545

[52] Li Y, Hu P, Liu Z, Liu X, Wei F, Zhou H, et al. Favorable neurological recovery for MESCC-induced paralysis with median 9-day duration before surgery. Journal of Orthopaedic Surgery and Research 2024;**19**:774

[53] Laufer I, Zuckerman SL, Bird JE, Bilsky MH, Lazáry Á, Quraishi NA, et al. Predicting neurologic recovery after surgery in patients with deficits secondary to MESCC: Systematic review. Spine. 2016;**41**(Suppl 20): S224–30

[54] Feins S, Kong W, Williams EF, Milone MC, Fraietta JA. An introduction to chimeric antigen receptor (CAR) T-cell immunotherapy for human cancer. American Journal of Hematology. 2019;**94**(S1):S3–S9

[55] Karschnia P, Jordan JT, Forst DA, Arrillaga-Romany IC, Batchelor TT, Baehring JM, et al. Clinical presentation, management, and biomarkers of neurotoxicity after adoptive immunotherapy with CAR T cells. Blood. 2019;**133**:2212–2221

[56] Karschnia P, Dietrich J. Neurological complications of CAR T cell therapy for cancers. Nature Reviews Neurology. 2025;**21**(8):422–431

[57] Butt OH, Zhou AY, Caimi PF, Luckett PH, Wisch JK, Derenoncourt PR, et al. Assessment of pretreatment and posttreatment evolution of neurofilament light chain levels in patients who develop immune effector cell–associated neurotoxicity syndrome. JAMA Oncology. 2022;**8**:1652

[58] Santomasso BD, Gust J, Perna F. How I treat unique and difficult to manage cases of CAR T-cell therapy associated neurotoxicity. Blood. 2023;**141**(20):2443–2451

[59] Favero A, Mohamed S, Battista ML, Cenacchi V, Sartori A, Mazzon G, et al. Immune effector cell–associated neurotoxicity delayed relapse after chimeric antigen receptor t-cell therapy. Neurology Neuroimmunology & Neuroinflammation. 2026;**13**(1): e200515

[60] Jain MD, Smith M, Shah NN. How I Treat Refractory CRS and ICANS Following CAR T-cell Therapy. Blood. 2023;**141**(20):2430–2442

[61] Gomez-Llobell M, Serra Smith C, Gómez-Costas D, Gómez-Centurión I, García Domínguez JM, Escudero-Vilaplana V, et al. ICANS risk model in CD19 CAR-T therapy: Insights from serum and CSF cytokine profiling. Bone Marrow Transplantation. 2025;**60**: 1351–1360

[62] Yu SL, Abdulhaleem M, Ul Abedin MS, Holley N, Al-saeed N, Anaum S, et al. The effect of statin use on outcome in patients receiving chimeric antigen receptor t-cell therapy. Transplantation and Cellular Therapy. 2025;**31**:S247

[63] Rees JH. Management of Immune Effector Cell-Associated Neurotoxicity Syndrome (ICANS). In: Kroger N, Gribben J, Chabannon C, Ibrahim YA, Einsele H, eds.The EBMT/EHA CAR-T Cell Handbook. Cham: Springer International Publishing; 2022. p. 141–145. DOI: 10.1007/978-3-030-94353-0_27

[64] Garcia Borrega J, Gödel P, Rüger MA, Öa O, Shimabukuro-Vornhagen A, Kochanek M, et al. In the Eye of the Storm: Immune-mediated Toxicities Associated With CAR-T Cell Therapy. HemaSphere. 2019;**3**(2):e191

Chapter 4

Prognostic Evaluation of Children with Brain Tumors Associated with Obstructive Hydrocephalus

Angela Leanca and Valeriu Timirgaz

Abstract

With the increase in morbidity and the number of people with neurological disabilities, neurosurgeons pay great attention to brain tumors and obstructive hydrocephalus. The incidence of secondary hydrocephalus in brain tumors in children can reach 73–90% of cases and is determined by the localization of neoplasms in the posterior fossa, with the involvement of the ventricular system. Treatment of concomitant hydrocephalus in such patients is closely related to the restoration of cerebrospinal fluid circulation up to the stage of surgical treatment of the tumor itself, thus representing a complex task. The purpose of the research is optimizing the surgical treatment of children with obstructive hydrocephalus associated with brain tumors by implementing and improving cerebrospinal fluid diversion methods. Thesis goals are: analysis of clinical and diagnostic features of children with an established diagnosis of obstructive hydrocephalus in brain tumors; Evaluation of surgical treatment in children with brain tumors and obstructive hydrocephalus. The paper represents a descriptive study of the diagnosis and surgical treatment of 88 children up to 18 years of age with intracerebral tumors and obstructive hydrocephalus, who required surgical interventions to restore CSF circulation by applying the following methods of diversion: DVE, VCS, ventriculoperitoneal shunt, and tumor ablation. In conclusion, persistent hydrocephalus is not dependent on primary CSF diversion. Evaluation of surgical treatment established non-obstructive hydrocephalus after tumor ablation in both patients in group L1 and children in group L0 ($p > 0.05$). The advantages of primary CSF diversion were demonstrated by a reduction in surgical time from 1 hour to 3 hours ($p < 0.001$) and intraoperative hemorrhage respectively ($p < 0.01$).

Keywords: brain, tumors, obstructive, hydrocephalus, prognostic

1. Introduction

Ventriculoperitoneal shunting plays a major role in the treatment of hydrocephalus of any etiology. Since the first shunt implantation by Nulsen about 50 years ago, a wide variety of shunt systems have been developed to provide safe and long-term control of hydrocephalus, but the overall high risk of complications, which can reach 50%, is most often determined by the incidence of shunt dysfunction. Such a high complication rate, according to some authors, requires the expansion of indications for alternative methods of treating hydrocephalus, in particular, performing endoscopic ventriculostomy of the floor of the third ventricle (VCS) for obstructive forms of hydrocephalus. Although surgical management of posterior fossa tumors by tumor removal can restore CSF circulation, 10–30% of patients tend to present with persistent hydrocephalus after tumor resection. The effectiveness of any method of treating hydrocephalus is usually assessed by the duration of hydrocephalus control: from the moment of shunt or VCS implantation until the moment when reintervention is needed due to shunt or stoma dysfunction.

Other neurosurgeons prefer shunting, even in situations where VCS can be performed, on the grounds that the long-term benefit of the endoscopic approach over shunts is unknown, and the intraoperative complications of VCS are generally greater than those of traditional shunting.

Evaluating the effectiveness of shunting operations in children with hydrocephalus associated with brain tumors is a rather complex task due to the large number of specific factors that can influence the effectiveness of the operations: external drainage of cerebrospinal fluid, the rate of tumor progression, and the presence of additional operations for the tumor. The rational and efficient sequence of these stages in the Republic of Moldova, as well as the results of these interventions, has not been scientifically elucidated.

2. Literature review

Neoplasms of the central nervous system are common in the pediatric population and represent approximately 20% of all tumors.

[1–3]. Brain tumors have been ranked as the second most common malignant pathology in children aged 0–14 years globally, with an incidence ranging from 1.12 to 5.14 cases per 100,000 people, as well as locally. More than 4,000 new cases of central nervous system tumors in children are diagnosed annually in the United States. Intracerebral tumors in children are primary tumors, which have a cerebral structural origin: brain, meninges, pituitary gland, and skull.

The annual incidence of primary brain tumors in a separate group of children and adolescents in Europe is 2.99% [4, 5]. More than 4000 new cases of central nervous system tumors in children are diagnosed annually in the United States, representing 2–5 to 6.06 cases per 100,000 children [6].

Brain tumors were ranked as the second most common malignancy in children aged 0–14 years globally, with an incidence ranging from 1.12–5.14 cases per 100,000 persons, as well as locally. In the Philippines, the most common brain tumors were astrocytoma (25.8%), followed by medulloblastoma (23.9%) [7]. Approximately 6,500 cases of primary central nervous system (CNS) tumors were recorded annually in England and Wales between 1995 and 2000, in those aged 15 years and over, of which 58% were diagnosed with "malignant" histological forms, possibly residual embryonic tissue.

The spectrum of brain tumors in children differs in location, histology, and prognosis compared with adults, suggesting that the pathogenetic mechanisms are different. Supratentorial tumors predominate in the first two years of life, while infratentorial lesions are more common in the first decade of life; supratentorial tumors again predominate in late adolescence and adulthood [8].

In this context, the particularities of the central nervous system in young children (0–1 years) should be mentioned – plasticity and the presence of the fontanelle, which represent important pathophysiological factors in the appearance of the following changes during the child's objective examination:

- increase in the size of the neurocranium, with the development of a disproportion between the neurocranium and the facial skull (being noticed by the mother 1–2 months after birth), macrocephaly.

- bulging forehead, fontanelle growth, which can reach 8–10 cm in diameter and the tense shape of the fontanelle.

- "sunset" eyes

- among the suggestive signs for HO is the palpation of suture dehiscence,

- development of a prominent subcutaneous vascular network of the scalp.

For the group of children aged 1–5 years, the clinical manifestations are representative, largely dependent on the child's psycho-neurological development (speech, motor). However, among the neurological signs of impairment, we list gait and balance disorders, behavioral disorders, vision disorders, morning headaches associated with vomiting, primary convulsions, drowsiness, and apathy.

Brain imaging can often be an unreliable guide for detecting tumors and hydrocephalus in the early stages. Thus, when performing brain CT without a contrast medium, tumors having practically the same density as the brain substance are not distinctive.

The ophthalmological examination can assess:

I. *Degrees of optic nerve damage*:

   - venous congestion

   - papillary edema, which would be one of the most important signs indicating the development of hydrocephalus and a critical determinant in choosing the treatment strategy [128, 195]. Early signs of the development of papillary edema in children would be scotomas and dyschromatopsia.

   - papillary stasis, representing an advanced stage of optic nerve damage.

II. *Visual field*.

III. *The onset of diplopia,* would be an indicator of compression of the abducens nerve.

The results of another group of researchers on a group of 419 children up to 15 years of age, over a 10-year period (2005–2015), established an incidence of papilledema in 14%–50% of children with shunt dysfunction and approximately 65% in children with primary intracerebral tumors. From this group, 46 patients remained eligible, according to the inclusion criteria, and were included in this study. The mean age was 6.3 ± 4.7 years. Nineteen patients who were not diagnosed with papilledema were identified; their mean age was 2.7 ± 2.7 years, and the other 27 patients, with papilledema established during the ophthalmological examination, had a mean age of 8.8 ± 4.2 years ($p < 0.001$).

Diagnostic imaging is essential for the evaluation of brain tumors associated with obstructive hydrocephalus, from the diagnostic and surgical planning stage to the postoperative surveillance stage. Thus, it is important to mention that the preoperative imaging examination has the task of assessing the state of the brain structures in order to plan the volume of the surgical intervention. The preoperative examination algorithm must include the entire spectrum of imaging investigations (MRI, CT, cerebral ultrasonography).

Ultrasound or neurosonography is still practiced today transfontanellar in cases where the fontanelle is open. It is a non-invasive method that allows the study of dilated ventricles and possible associated anomalies (calcifications, cysts). The advantage of this method is that it can be repeated whenever necessary. USG examination in children can be used as a screening method for selecting patients for CT or MRI, with MRI being preferable. The MRI method has a significantly higher sensitivity than other diagnostic methods in the detection and dynamic evaluation of both the given pathology and other concomitant pathologies, including congenital ones. Magnetic resonance imaging (MRI) is the gold standard when it comes to the evaluation of brain tumors [9]. The standard protocol, at the preoperative stage, includes pre – and post-contrast sequences in the T1, T2, and FLAIR regimes. Diffusion tensor imaging (DTI) can help assess the location and integrity of white matter tracts at the preoperative stage. Postoperatively, to assess the state of brain structures, the presence or absence of hydrocephalus, cerebral MRI with flowmetry is indicated [10].

Research into molecular mechanisms underlies the assessment of the causes of intracerebral tumors in children, and ongoing research has established the classification of different tumor types into several subgroups, based on the genomic profile, establishing correlations between the specificity of different combinations of genomic alterations and subsequent clinical outcomes. Molecular subclassification was integrated into the fifth edition of the World Health Organization (WHO) Classification of Central Nervous System Tumors, which was published in 2021, with the grading of certain tumors based on genomic alterations.

Several molecular methods are known to be used in the diagnosis of tumor formation and prognostic assessment:

- FISH – fluorescent in situ hybridization method
- PCR – polymerase chain reaction
- NGS – next-generation Sanger sequencing

Surgical ablation of intracerebral tumors in children is the primary method of choice, as it represents the direct removal of the pathological process, which is absolutely necessary for establishing the definitive diagnosis and choosing adjuvant treatment in the postoperative stage, as well as resolving obstructive hydrocephalus [11, 12]. The stages of surgical treatment of brain tumors in children need to be carefully evaluated in consultation, starting from the fact that most intracerebral tumors in children require repeated surgical treatment, determined by the presence of relapse.

In most cases, according to contemporary requirements, for certain stages of the ablation operation of a brain tumor, an operating microscope is used, which has certain priorities: a small distance between the objective and the eyepieces, variable focus and zoom, binocular view, presence of a varioscope, Xenon lamps, and video recording. Because in cases of deeply located tumors or those adjacent to eloquent regions, only a fine technique can ensure optimal visual control of surgical manipulations.

There are two fundamentally different methods of brain tumor resection: eccentric and circumferential, with delimitation at the periphery and tissue aspiration in the perifocal region of the tumor. If possible, the second method is preferable. Unfortunately, delimitation of the tumor in the region of perifocal edema is possible only in tumors with nodular growth, when the boundary between the tumor and brain tissue is clear. If the neoplasm diffusely affects the brain tissue in the eloquent regions, the degree of surgical radicality is decided intraoperatively, taking into account the major risk of invalidating the patients.

The most common histological forms in children aged 0–9 years include gliomas and embryonal tumors; in children aged 10–14 years, gliomas and pituitary tumors are the most common tumor types. In adolescents aged 15–19 years, pituitary tumors and gliomas again predominate [13].

Gliomas are considered to be the most common tumor subtype (46.2%), followed by meningiomas (25.6%), consistent with studies conducted in Ghana, India, and the United States. Low-grade gliomas (WHO grades I and II) are a heterogeneous group, constituting the largest subgroup of pediatric brain tumors. The primary cell types of these tumors are diverse and include neuronal, mixed glioneuronal, astrocytic, oligodendroglial, or oligoastrocytic cell types. Often, low-grade pediatric gliomas are caused by alterations in the mitogen-activated protein kinase (MAPK) signaling pathway, including BRAF mutation or fusion, FGFR1 mutation or structural rearrangement, NF1 mutation, and RAF and NTRK family fusions.

Consistent with other studies, a male predominance (65%) has been found in patients diagnosed with gliomas, and the majority of cases (80%) occur in adults. In this study, craniopharyngioma (27.3%) was the most common tumor diagnosed in children and adolescents, followed by medulloblastoma, glioblastoma, and gliomas (18.2% each). Various pediatric studies have reported different rates of occurrence for different tumor subtypes. These differences have been hypothesized to be influenced by viral and bacterial infections, ionizing radiation, and other environmental factors [14].

Choroid plexus tumors are rare neuroepithelial tumors with a predilection for intraventricular locations, accounting for less than 1% of all intracranial tumors and less than 4% of brain tumors in children aged 1–3 years. Choroid plexus papilloma (CPP) is a rare benign tumor arising from the choroidal villi. Pediatric patients usually present with acute signs of hydrocephalus, requiring immediate surgical intervention. They are usually located in the fourth ventricle in adults and the lateral ventricle in infants and children. The third ventricle is a very rare location, with only a few cases reported in the literature [15]. The challenges associated with these lesions are related to the flow of cerebrospinal fluid and the intense vascularity, which can lead to significant blood loss during surgery, necessitating incomplete resection of these tumors [16].

Although it is a common cause of mortality globally, there is a low survival rate in low – and middle-income populations, contrary to a high survival rate in high-income populations, regardless of advances in the diagnosis and surgical treatment of tumors applied to children [17–19].

Ventriculoperitoneal shunting plays a major role in the treatment of hydrocephalus of any etiology. Since the first shunt implantation by Nulsen about 50 years ago, a wide variety of shunt systems have been developed to provide safe and long-term control of hydrocephalus, but the overall high risk of complications, which can reach 50%, is most often determined by the incidence of shunt dysfunction. Such a high complication rate, according to some authors, requires the expansion of indications for alternative methods of treating hydrocephalus, in particular, performing endoscopic ventriculostomy of the floor of the third ventricle (VCS) for obstructive forms of hydrocephalus.

Other neurosurgeons prefer shunting, even in situations where VCS can be performed, on the grounds that the long-term benefit of the endoscopic approach over shunts is unknown, and the intraoperative complications of VCS are generally greater than those of traditional shunting.

According to literature data, posterior fossa tumors represent 7.9% of intracranial lesions, and approximately 20–90% of patients with posterior fossa tumors present with obstructive hydrocephalus. Although surgical management of posterior fossa tumors by tumor removal can restore CSF circulation, 10–30% of patients tend to present with persistent hydrocephalus after tumor resection. The effectiveness of any method of treating hydrocephalus is usually assessed by the duration of hydrocephalus control: from the moment of shunt or VCS implantation until the moment when reintervention is needed due to shunt or stoma dysfunction. Evaluating the effectiveness of shunting operations in children with hydrocephalus associated with brain tumors is a rather complex task due to the large number of specific factors that can influence the effectiveness of the operations: external drainage of cerebrospinal fluid, the rate of tumor progression, and the presence of additional operations for the tumor. The rational and efficient sequence of these stages in the Republic of Moldova, as well as the results of these interventions, is not scientifically elucidated.

3. Material and methods

3.1 Study design and overall procedure

The paper presents a descriptive study of the diagnosis and surgical treatment of 88 children up to 18 years of age with intracerebral tumors and obstructive hydrocephalus, who required surgical interventions to restore CSF circulation by applying the following methods of diversion: DVE, VCS, ventriculoperitoneal shunt, and tumor ablation.

In order to develop the doctoral thesis, the positive opinion of the Research Ethics Committee of the "Nicolae Testemițanu" State University of Medicine and Pharmacy was obtained (Minutes no. 76 of November 14, 2016).

3.2 Survey and data collection

All interventions were performed at the Institute of Mother and Child and SCMC "V. Ignatenco" during the period 2013–2018 (5 years). The following research methods were used: (1) clinical investigation (objective examination, neurological examination); (2) laboratory methods (blood analysis, blood biochemical analysis, CSF biochemical analysis); (3) imaging investigation methods (chest X-ray, panoramic abdominal and pelvic X-ray, NSG, brain CT, brain MRI with contrast); (4) functional examination methods (ophthalmoscopy, EEG); (5) treatment methods for obstructive hydrocephalus and intracerebral tumors; (6) morphopathological methods (histology, immunohistochemistry). These data were statistically processed, and the survival rate was established according to Kaplan-Meier. The representation of the results of the statistical data is shown using graphs. Preoperative and postoperative information, obtained based on the analysis of all available information contained in the paper versions of the disease history from the archives of the institutions, was recorded in the database.

The allocation of the child to one of the research groups was performed post factum, according to the establishment of clinical, diagnostic, and treatment criteria surgically applied for obstructive hydrocephalus within the brain tumor, confirmed by imaging examinations, medical history, physical, and neurological examination.

L0 group inclusion criteria:

1. Child aged 0–17 years with an established diagnosis of an intracerebral tumor associated with obstructive hydrocephalus, as identified through imaging investigations by CT/MRI of the brain.

2. Child has not undergone CSF diversion surgery.

3. Child under the care of a pediatric oncologist in the postoperative stage.

L0 group inclusion criteria:

1. Child aged 0–17 years, with an established diagnosis of intracerebral tumor associated with obstructive hydrocephalus, confirmed through imaging investigations by CT/MRI of the brain.

2. The child has undergone surgery with CSF diversion.

3. The child is on the record of the pediatric oncologist at the postoperative stage.

Exclusion criteria:

1. Children older than 17 years.

2. Children with a diagnosis of intracerebral tumor without obstructive hydrocephalus.

3. Children with an established diagnosis of an intracerebral tumor associated with obstructive hydrocephalus on imaging investigations by CT/MRI of the

brain, but who have not undergone surgery in the neurosurgery department of the Institute of Mother and Child and "V. Ignatenco".

We considered those situations in which patients, in addition to CSF diversion surgery, underwent other types of surgery, for example, removal of a recurrent tumor or biopsy. In these cases, the time elapsed from shunt implantation to the time of the first shunt revision was assessed. The form of hydrocephalus was determined based on neuroimaging data. If there were signs indicating the presence of hydrocephalus and a brain tumor, hydrocephalus was defined as occlusive. In the presence of symmetrical dilatation of the entire ventricular system and/or subarachnoid spaces, in the absence of obvious signs of tumor occlusion, hydrocephalus was defined as non-obstructive, and indications for ventriculoperitoneal shunting operations for non-obstructive forms of hydrocephalus were assessed.

The histology and localization of the tumor were taken into account. In the groups of patients with a confirmed histological diagnosis, patients were stratified depending on the degree of malignancy of the brain tumor.

In the case of a ventriculoperitoneal shunt, the position of the catheter in the ventricular system (anterior horn, posterior horn, bilateral implantation) was taken into account, as well as the presence of external ventricular drainage before shunting and the duration of external drainage placement.

Early postoperative outcomes were assessed at the time of discharge based on the analysis of medical history. The main sources of information on the causes of shunt dysfunction were follow-up data obtained from the medical records of repeated inpatient admissions.

3.3 Evaluation and discussion

Socio-demographic and anamnestic data. In the general study group, 88 children aged 0–17 years were selected, including 41 (46.6%; 95% CI: 36.4–57.0) male patients and 47 (53.4%; 95% CI: 43.0–63.6) female patients (**Figure 1**):

In this figure, a tendency to affect male children in comparison to female children is observed.

Depending on the area of residence, 32 (36.4%; 95% CI: 26.9–46.7) patients lived in the northern area of the Republic of Moldova, 39 (44.3%; 95% CI: 34.3–54.7) patients lived in the central area of the Republic of Moldova, and 17 (19.3%; 95% CI: 12.1–28.5) patients lived in the southern area of the Republic of Moldova (**Figure 2**):

This figure shows the increased incidence in downtown areas, determined by access to high-performance investigations in specialized centers.

Analysis of the age of patients at the time of hospitalization found that 7 (**8.0%;** 95% CI: 3.6–15.0) children were aged from 0 to 1 year, 32 (**36.4%;** 95% CI: 26.9–46.7) children – from 1 year to 5 years, 23 (**26.1%;** 95% CI:17.8–36.0) children – from 5 years to 10 years and 26 (**29.5%;** 95% CI: 20.8–39.6) children – from 10 years to 17 years (**Figure 3**):

The figure shows the predominance of children aged 1–5 and 10–17 years.

In the patients in our study group, the disease had an *acute onset* in 25 (28.4%; 95% CI: 19.8–38.4) cases, and a delayed onset was observed in 63 (71.6%; 95% CI: 61.6–80.2).

Due to the *duration of the anamnesis* and the analysis of the period from the appearance of the first symptoms, as attested by the parents, to the hospitalization

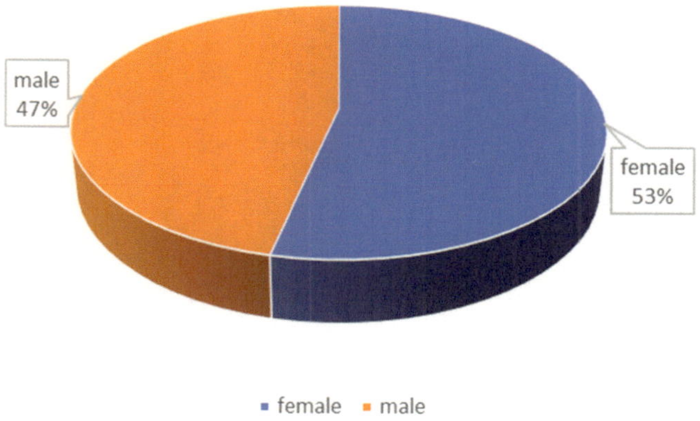

Figure 1.
Distribution of patients with intracerebral tumors associated with obstructive hydrocephalus by sex (%).

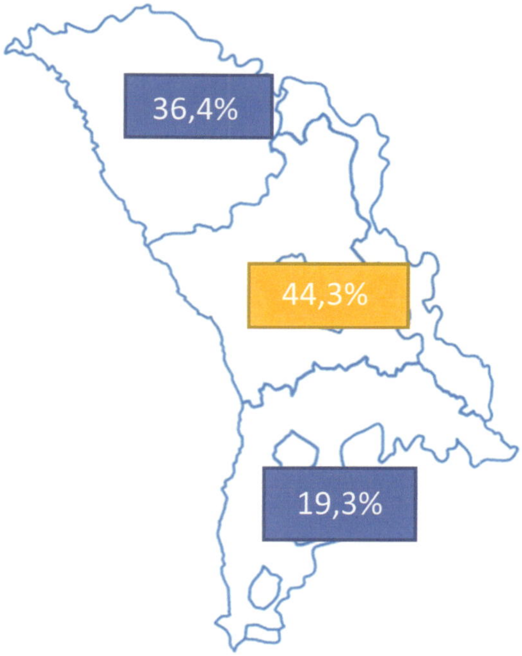

Figure 2.
Distribution of patients by area of residence in the Republic of Moldova.

for the diagnosis of an intracerebral tumor associated with obstructive hydrocephalus, was as follows (**Figure 4**):

- with the highest frequency were patients with a duration from 0 to 3 months, in total 50 (56.8%; 95% CI: 46.4–66.8) patients were hospitalized,
- from 3 to 6 months – 18 (20.5%; 95% CI: 13.1–29.8) patients,
- from 6 to 9 months – 3 (3.4%) patients,

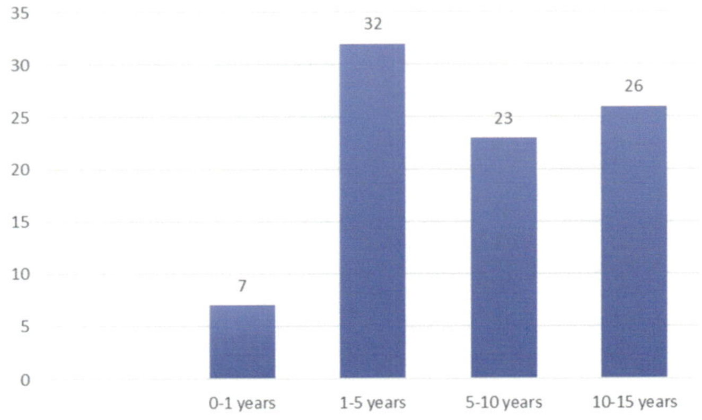

Figure 3.
Distribution of patients with intracerebral tumors associated with obstructive hydrocephalus according to age (%).

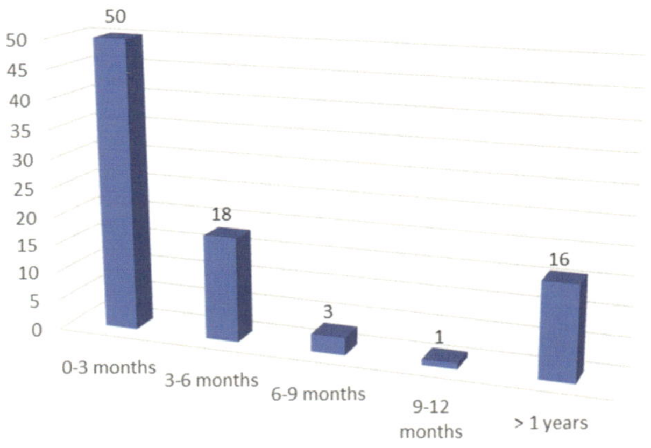

Figure 4.
Distribution of patients with brain tumors associated with obstructive hydrocephalus according to the period from the appearance of the first symptoms to hospitalization (%).

- from 9 to 12 months – 1 (1.1%; 95% CI: 0.1–5.2) patient and from 12 months or more – 16 (18.2%) patients).

In this research, I found out that the prediagnosis period was 60 days.

The L0 group includes patients treated for brain tumor resection without cerebrospinal fluid diversion. Depending on the degree of tumor resection, the following distribution was determined:

- *total ablation* was performed in 30 (68.2%; 95% CI: 53.6–80.5) patients who had well-demarcated tumors with clear margins, associated with cysts, non-infiltrative, and without areas of hemorrhage or ischemia, such as astrocytomas.

- *subtotal resection* – in 14 (31.8%; 95% CI: 19.5–46.4) patients, determined by adherent, infiltrative tumors located in the IV ventricle and rhomboid fossa, with hemorrhage from the tumors, such as ependymomas, choroid plexus papillomas, and medulloblastomas.

Based on the intraoperative particularities, the duration of intracerebral tumor ablation was:

- 1–3 hours in 2 (4.5%; 95% CI: 1.0–13.8) cases,
- 3–6 hours in 37 (84.1%; 95% CI: 71.3–92.6) cases
- 6–9 hours in 5 (11.4%; 95% CI: 4.5–23.1) cases.

Children who underwent surgery for a period longer than 3 hours and subtotal ablation were identified in 12 (27.3%; 95% CI: 15.9–41.6) patients from group L0, as determined by the intraoperative particularities:

- lack of distinctive markings between healthy tissue and tumor,
- adherent and infiltrative tumor,
- hemorrhage from tumors.

In the early postoperative period (until discharge from the hospital), in 27 (61.4%; 95% CI: 46.6–74.7) patients were found: malignant hyperthermia in 2 (4.5%; 95% CI: 1.0–13.8) patients, neurological disorders in 17 (38.6%; 95% CI: 25.3–53.4) patients persistent active hydrocephalus in 8 (18.2%; 95% CI: 9.0–31.4) patients, who were surgically intervened by applying ventriculoperitoneal shunt in 6 children, and in 2 children endoscopic VCS was performed, pseudomeningocele in 4 (9.1%; 95% CI: 3.1–20.2) children.

The definitive postoperative diagnosis of primary tumors, established at histological examination, included: medulloblastoma – in 8 (18.2%; 95% CI: 9.0–31.4) patients, ependymoma – in 11 (25.0%; 95% CI: 14.1–39.1) patients, astrocytoma – in 12 (27.3%; 95% CI: 15.9–41.6) patients, glioblastoma – in 2 (4.5%; 95% CI: 1.0–13.8) patients, neuroblastoma – in 2 (4.5%; 95% CI: 1.0–13.8) patients, craniopharyngioma – in 4 (9.1%; 95% CI: 3.1–20.2) patients, and choroid papilloma – in 5 (11.4%; 95% CI: 4.5–23.1) patients.

The L1 group includes patients treated for brain tumor resection and cerebrospinal fluid diversion. In the group of L1 patients, depending on the degree of tumor resection, the following types of primary tumor ablation were applied (**Figure 5**): total ablation was performed in 32 (72.7%; 95% CI: 58.4–84.1) patients,

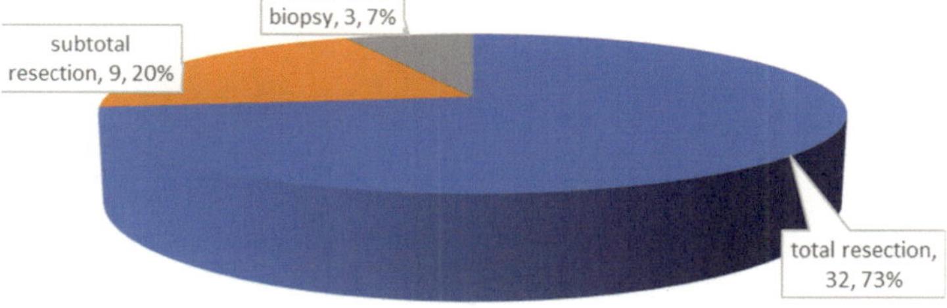

Figure 5.
Types of tumor resection in children from L1 group.

subtotal ablation in 9 (20.5%; 95% CI: 10.6–34.0) patients, and biopsy was performed in 3 (6.8%; 95% CI: 2.0–17.1) patients.

In this figure, we can see that total ablation was performed in 32 (72.7%; 95% CI: 58.4–84.1) patients.

Repeated hospitalization, in the late postoperative period up to 36 months of follow-up, required 21 (47.7%; 95% CI: 33.5–62.2) patients.

The main causes of repeated hospitalization were:

- Non-occlusive hydrocephalus determined by shunt dysfunction in 7 (15.9%; 95% CI: 7.4–28.7) cases.

- Recurrent tumor with hydrocephalus in 12 (27.3%; 95% CI: 15.9–41.6) patients, of which 6 children required de novo application of the V–P shunt, as an additional surgical intervention after tumor ablation.

- Recurrent tumor without hydrocephalus in 2 (4.5%; 95% CI: 1.0–13.8) patients.

Tumor recurrence was diagnosed in 17 (38.6%; 95% CI: 25.3–53.4) patients, including localization in the same region in 11 (25.0%; 95% CI: 14.1–39.1) patients and another localization in 6 (13.6%; 95% CI: 5.9–26.0) patients.

3.4 The prognostic evidence

Following the analysis of the survival period at the early and late postoperative stages, using the Kaplan-Meyer method, a statistically insignificant difference was established in both groups.

Survival duration according to the period of months was similar in both study groups and is shown in **Figure 6**:

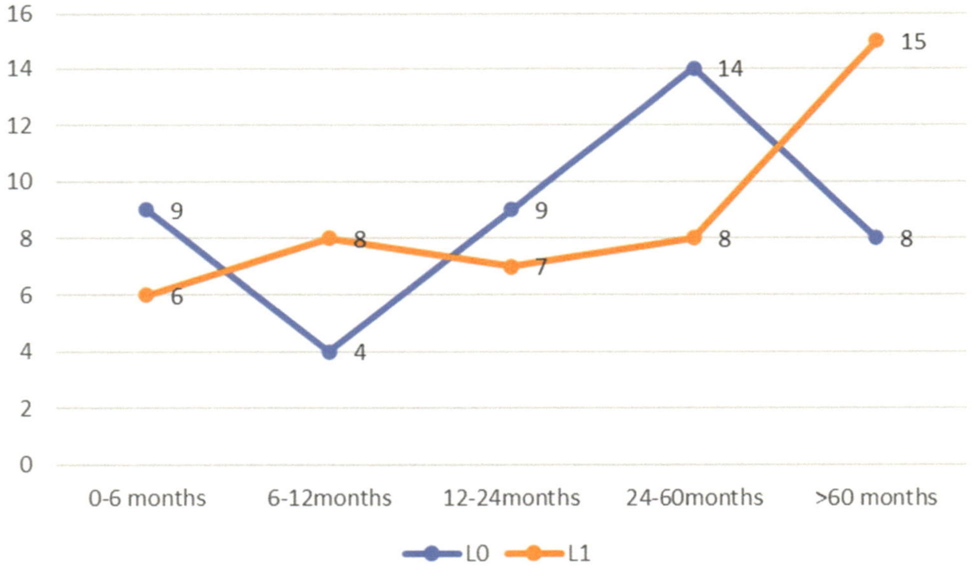

Figure 6.
Survival of patients with intracerebral tumors associated with obstructive hydrocephalus from both study groups (days).

- from 0 months to 6 months 6 (13.6%; 95% CI: 5.9–26.0) patients from group L1 and 9 (20.5%; 95% CI: 10.6–34.0) patients from group L0 survived ($p > 0.05$),
- from 6 months to 12 months – 8 (18.2%; 95% CI: 9.0–31.4) patients from group L1 and 4 (9.1%; 95% CI: 3.1–20.2) patients from group L0 ($p > 0.05$),
- from 12 months to 24 months – 7 (15.9%; 95% CI: 7.4–28.7) patients from group L1 and 9 (20.5%; 95% CI: 10.6–34.0) patients from group L0 ($p > 0.05$),
- from 24 months to 60 months – 8 (18.2%; 95% CI: 9.0–31.4) patients from group L1 and 14 (31.8%; 95% CI: 19.5–46.4) patients from group L0 ($p > 0.05$),
- from 60 months or more – 15 (34.1%; 95% CI: 21.4–48.8) patients from group L1 and 8 (18.2%; 95% CI: 9.0–31.4) patients from group L0 ($p > 0.05$).

Compared with patients with intracerebral tumors associated with obstructive hydrocephalus from group L1, the average survival duration showed only an increasing trend in the group of patients from group L0 (813.80 ± 1012.7 days; Md – 210.0, IIQ: 30.0–1617.5 and 752.30 ± 1009.2 days; Md – 180.0, IIQ: 30.0–1117.5, respectively; $p > 0.05$).

4. Conclusion

Persistent non-obstructive hydrocephalus after tumor ablation was established in 5 (11.4%; 95% CI: 4.5–23.1) patients from group L1 and in 8 (18.2%; 95% CI: 9.0–31.4) children from group L0 ($p > 0.05$), that's why the evaluation of children with brain tumors associated with obstructive hydrocephalus and the establishment, in a multidisciplinary board, of clinical-neurological and imaging criteria, represents the basic stage in the assessment of the differentiated treatment algorithm.

Conflict of Interest

The authors have no conflict of interests to declare that are relevant to the content of this article.

Competing Interests

The authors declare that they have no competing interests.

Notes

Sincere thanks to the thesis supervisor, Valeriu Timirgaz, for guidance throughout the research.

Appendices and nomenclature

MRI	Magnetic resonance imaging
CT	Computed tomography
CSF	Cerebrospinal fluid
VCS	Ventriculocisternostomy
VPS	Ventriculoperitoneal shunt
EVD	External ventricular drainage
VAS	Ventriculoatrial shunt
EEG	Electroencephalography
USG	Ultrasound
NSG	Neurosonography
FO	Fundus of the eye
PNO	Optic nerve papilla
CNS	Central nervous system
HO	Obstructive hydrocephalus
L1	Children with intracerebral tumors associated with obstructive hydrocephalus treated by CSF diversion and tumor ablation
L0	Children with intracerebral tumors associated with obstructive hydrocephalus treated by tumor ablation without CSF diversion

Author details

Angela Leanca[1*] and Valeriu Timirgaz[2]

1 Child Municipal Hospital "Valentin Ignatenco", Chișinău, Republic of Moldova

2 The Institute of Neurology and Neurosurgery "Diomid Gherman," Chișinău, Republic of Moldova

*Address all correspondence to: leanca.angela@mail.ru

References

[1] Boethun A, Vissing NH, Foss-Skiftesvik J. CNS infection in children with brain tumors: Adding ventriculostomy to brain tumor resection increases risk more than 20-fold. Child's Nervous System. 2023;**39**:387–394. DOI: 10.1007/s00381-022-05799-8

[2] Brown CE, Bucktrout S, Butterfield LH, Futer O, Galanis E, Hormigo A, Lim M, Okada H, Prins R, Marr SS, Tanner K. The future of cancer immunotherapy for brain tumors: A collaborative workshop. Journal of Translational Medicine. 2022;**20**. DOI: 10.1186/s12967-022-03438-z

[3] Cornejo P, Feygin T, Vaughn J, CM P, Korostyshevska A, Patel M, Dme B, Miller J, Goncalves LF. Imaging of fetal brain tumors. Pediatric Radiology. 2020;**50**:1959–1973. DOI: 10.1007/s00247-020-04937-1

[4] Edgeworth J, Bullock P, Bailey A, Gallagher A, Crouchman M. Why are brain tumours still being missed? Archives of Disease in Childhood. 1996;**74**:148–151. DOI: 10.1136/adc.74.2.148

[5] Lacour B, Guyot-Goubin A. Incidence of childhood cancer in France. European Journal of Cancer Prevention. **19**;173-181. DOI: 10.1097/CEJ.0b013e32833876c0

[6] Roth J, Constantini S. Neurofibromatosis type 1–related hydrocephalus: causes and treatment considerations. Neurology India. 2020;**36**(Suppl 2):2385–2390. DOI: 10.1007/s00381-020-04719-y

[7] Montoya JEM, Moran MAM, Rodriguez PS, et al. Choroid plexus papilloma of the fourth ventricle: A pediatric patient. Asian Journal of Neurosurgery. 2019;**14**(2):585–588. DOI: 10.4103/ajns.AJNS_301_18

[8] Orduna PC, Cheryl Anne PL-S. Factors associated with delayed diagnosis among Filipino pediatric brain tumor patients: A retrospective review. CNS Oncology. 2022;**11**. DOI: 10.2217/cns-2022-0009

[9] Pancaldi A, Pugliese M, Migliozzi C, Blom J, Cellini M, Iughetti L. Neuropsychological outcomes of children treated for brain tumors. Children. 2023;**10**:472. DOI: 10.3390/children10030472

[10] Dvir JSR, Ben-Sira L, Yalon M, Boop F, Constantini S, Roth J. MRI-based diagnosis and treatment of pediatric brain tumors: Is tissue sample always needed? Child's Nervous System. 2021;**37**:1449–1459. DOI: 10.1007/s00381-021-05148-1

[11] de Oliveira RS, Jucá CEB, Valera ET, Machado HR. Hydrocephalus in posterior fossa tumors in children. Are there factors that determine a need for permanent cerebrospinal fluid diversion? Child's Nervous System. 2008;**24**:1397–1403. DOI: 10.1007/S00381-008-0649-X

[12] Wong T-T, Liang M-L, Chen H-H, Chang F-C. Hydrocephalus with brain tumors in children. Child's Nervous System. 2011;**27**:1723–1734. DOI: 10.1007/s00381-011-1523-9

[13] Lacour B, Guyot-Goubin A, Guissou S, Bellec S, Desandes E, et al. Incidence of childhood cancer in France: National Children Cancer Registries,

2000-2004. European Journal of Cancer Prevention. 2010;**19**:173–181

[14] Klein-Geltink J, Pogany L, Mery LS, Barr RD, Greenberg ML. Impact of age and diagnosis on waiting times between important healthcare events among children 0 to 19 years cared in pediatric units: The Canadian Childhood Cancer Surveillance and Control Program. Journal of Pediatric Hematology/Oncology. 2006;**28**:433–439. DOI: 10.1097/01.mph.0000212945.20480.26

[15] Tania S-M, Shields M. Pediatric Brain Tumors: Diagnosis and Management. Journal of Pediatric Oncology Nursing. 1995;**12**(4):188–198. DOI: 10.1177/104345429501200403

[16] Bondy M, El-Zein R, Wrensch M. Epidemiology of Brain Tumors. In: Brain Tumors. 2005. p. 3–18. DOI: 10.1385/1-59259-843-9:003

[17] Christopher WMNGC. Advances in chimeric antigen receptor (CAR) T-cell therapies for the treatment of primary brain tumors. Antibodies. 2022. DOI: 10.3390/antib11020031

[18] Desjardins L, Thigpen JC, Kobritz M, Bettis AH, Gruhn MA, Ichinose M, Hoskinson K, Fraley C, Vreeland A, McNally C, Compas BE. Parent reports of children's working memory, coping, and emotional/behavioral adjustment in pediatric brain tumor patients: A pilot study. Child Neuropsychology. 2018;**24**:959–974. DOI: 10.1080/09297049.2017.1365828

[19] Gurney JG, Kadan-Lottick N. Brain and other central nervous system tumors: Rates, trends, and epidemiology. Current Opinion in Oncology. 2001;**13**:160–166. DOI: 10.1097/00001622-200105000-00005

Chapter 5

Traumatic Subaxial Cervical Facet Dislocation: Current Concepts in Pathophysiology, Diagnosis and Evidence-Based Management

Eduardo Bertolini

Abstract

Among traumatic spinal injuries, subaxial cervical trauma with facet joint dislocation is undoubtedly a prevalent pathology with a high risk of neurological sequelae. The literature is controversial regarding how to treat these cases and what the best surgical approach is. With the advent of Magnetic Resonance Imaging (MRI), the presence of traumatic disc herniations has become a point of concern regarding how to reduce these dislocations, whether it is necessary to treat them before reduction, and the possibility of neurological worsening with the use of cranial traction. Is closed reduction with traction mandatory before fixation surgery, or is it an unnecessary step considering that reduction is easily achieved during surgery? Would these points be better addressed through an anterior, posterior, or combined approach? If combined, what is the best sequence to perform the treatment without increasing the risk of neurological worsening? The objective of this chapter is to present the evidence from the literature on this topic, aiming to guide the spine surgeon's conduct in these cases. Mechanism of Injury, Pathophysiology, Clinical Presentation, radiological investigation (X-Ray, Computed Tomography, Magnetic Resonance Imaging), Traumatic Disc Herniation, Closed Reduction, Operative Management, Anterior Cervical Approach, Posterior Cervical Approach, Combined (360°) Approach, Triple Approach (Anterior–Posterior–Anterior), Timing of Intervention will be discussed. To obtain the best answers from the literature, ideally it should use tools from Evidence-Based Medicine, with analysis of systematic review articles, preferably with meta-analyses, including well-designed articles. Reviews attempt to establish the best evidence among the available information.

Keywords: cervical injury, facet dislocation, subaxial, cervical fracture dislocation, traumatic disc herniation

1. Introduction

Traumatic injuries of the cervical spine, especially those involving facet dislocation and spondylolisthesis, are among the most devastating pathologies that spine surgeons encounter, both due to surgical difficulties, neurological sequelae, and the societal costs associated with the rehabilitation of these individuals. They are generally caused by high-energy mechanisms such as traffic accidents, shallow-water diving, and falls from standing height. Hyperflexion, which may be associated with posterior distraction, generally leads to these unilateral or bilateral facet dislocations, accompanied by posterior ligamentous failure. The advancement of imaging exams, primarily CT and MRI, has facilitated the evolution of surgical techniques and reduction strategies for these locked listheses. Evidence from the literature supports the need for early decompression and mechanical stabilization. However, issues such as the necessity of cranio-cervical traction for the reduction of dislocation, the presence of traumatic disc herniation, the requirement for MRI to diagnose traumatic herniations, and determining the best surgical technique for correcting facet dislocation with realignment and mechanical stabilization remain topics of discussion among specialists. Systematic reviews and meta-analyses are important tools in the effort to uncover the current evidence addressing these questions.

2. Mechanism of injury and pathophysiology

Subaxial cervical fracture-dislocations with facet dislocation are most frequently caused by flexion and posterior distraction mechanisms. In unilateral dislocations, a rotational mechanism is usually associated. The facet of the superior vertebra slides over the entire facet of the inferior vertebra, completely losing contact between their articular surfaces, leading to anterior displacement of the cranial vertebra during the trauma mechanism, which cannot be reduced due to facet locking. The structures that may be involved in these injuries are the facet joint capsule, interspinous ligament, supraspinous ligament, ligamentum flavum, anterior longitudinal ligament, annulus fibrosus of the disc, in addition to the vertebra itself. The integrity of the spinal cord depends, in most cases, on the intensity of the trauma and the amount of vertebral displacement during this movement. Just as in degenerative spondylolisthesis of the lumbar region, in which a pseudo-disc protrusion can be observed, in these traumatic cervical injuries the disc may protrude into the vertebral canal with spinal cord compression. The presence of these protrusions may increase the neurological risk during reduction maneuvers for spinal realignment.

3. Clinical presentation

The clinical findings related to subaxial cervical trauma with facet dislocation are variable and must be investigated when there is a history of trauma involved. Facet dislocations present a high risk of neurological injury due to the associated spondylolisthesis.

The main findings are:

- Cervical pain (with or without radiation)
- Limitation of neck movement
- Neurological symptoms, such as:

- Paresthesias in upper limbs or lower limbs
- Sensory changes
- Strength changes (upper limbs and/or lower limbs)
- Sphincter changes

4. Radiological evaluation

In suspected traumatic cervical injuries, radiological investigation is fundamental for diagnosis or to rule out risk conditions. In the hospital setting, where initial care is performed after referral by rescue services with immobilization and a cervical collar, the rational use of resources (imaging exams) and standardization of care are fundamental to ensure patient safety. For the specific case of cervical spine trauma, there are two main algorithms described in the literature with the purpose of safely removing the cervical collar without performing radiographs: the National Emergency X-Radiography Utilization Study (NEXUS) [1] and the Canadian C-Spine Rule (CCR) [2]. NEXUS uses five main criteria: no midline cervical tenderness, no focal neurologic deficit, normal alertness, no intoxication, and no painful, distracting injury. CCR uses high- and low-risk factors and the ability to rotate the neck 45° bilaterally. Systematic reviews reveal that both are safe for this evaluation, with CCR being slightly superior [3].

4.1 X-ray

Generally, the exam is available in emergency units of all hospitals in the world, including simpler cities in less developed countries and also in the best hospitals in the world. It provides information that may establish the diagnosis of cervical traumatic injuries, especially facet dislocations.

When performed with proper technique (with good visualization from the base of the skull and cranio-cervical junction to T1, without shoulder overlap on lateral views), fracture diagnoses are usually sufficient, and, in the case of facet dislocations, the association of AP and lateral views allows identification of unilateral or bilateral dislocations, as shown in **Figure 1**.

The main findings that should be investigated are:

- Good alignment on lateral views
- Alignment and rotation on AP views
- Fractures
- Increased distance between spinous processes
- Facet dislocations
- Soft tissue swelling (tracheal deviation and other structures)

4.2 Computed tomography (CT)

Currently, it is the modality of choice for the investigation of spine injuries in polytraumatized patients in the emergency room. If this resource is available, X-ray may even be dispensed with. In addition to the findings described in X-ray, it allows some visualization of soft tissues such as the spinal cord and discs (although MRI is superior).

Thus, we have:

- Good alignment in sagittal reconstructions
- Alignment and rotations in coronal views
- Fractures
- Increased distance between spinous processes
- Facet dislocations
- Deviation of the trachea and other adjacent structures
- Disc herniations
- Spinal cord compressions

4.3 Magnetic resonance imaging (MRI)

An important complementary exam to CT, more sensitive for visualizing spinal cord compression, nerve roots, hematomas, and the intervertebral disc. It is fundamental for evaluating ligament ruptures, which indicate instability. Ligament injuries are often difficult to diagnose and are frequently classified as "suspected injury," which may change treatment planning. In cases of facet dislocation specifically, there is great controversy in the literature regarding reduction planning and the possibility of iatrogenic injury if a traumatic disc herniation is present. Reduction maneuvers could lead to spinal cord compression and neurological worsening at the time of the maneuver. This important discussion raises the need for performing cervical MRI in all cases of facet dislocation for planning these maneuvers as well as surgical techniques. The main MRI findings to be observed are:

- Alignment in sagittal views
- Alignment and rotations in coronal views
- Fractures
- Increased distance between spinous processes
- Edema in ligamentous regions
- Ligament ruptures

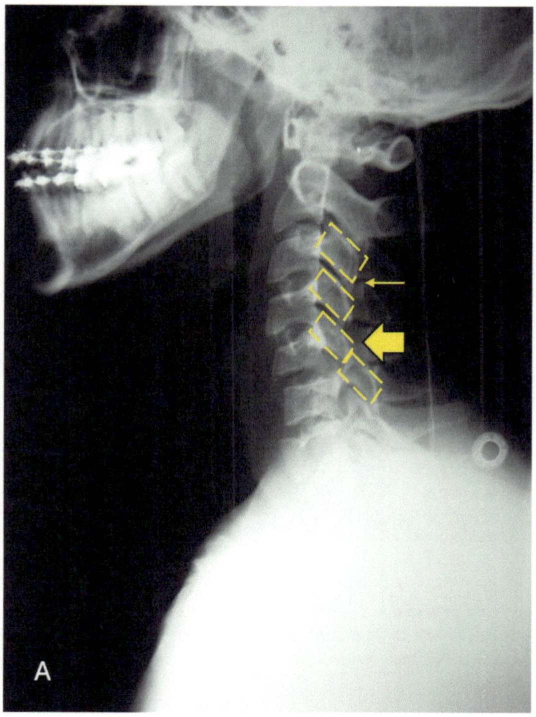

Figure 1.
X-ray, lateral view; traumatic spondylolisthesis C5/6 and facet dislocation; note normal facets C3/4 (thin arrow) and locked facets C5/6 (thick arrow: the articular surfaces of C5 and C6 have lost contact with each other.).

- Facet dislocations
- Hematomas
- Traumatic disc herniations
- Spinal cord compressions
- Nerve root compressions
- Spinal cord signal changes

5. Closed reduction

Closed reductions (**Figure 2**) using halo traction for cases of cervical traumatic injuries with facet dislocation were of utmost importance in the past and have become debatable nowadays with the advancement of surgical techniques and the availability of appropriate materials for these emergency procedures. Its application, which does not depend on the availability of an operating room or complex materials – only the halo – makes it a low-cost resource with easy and rapid installation that may unlock the facets, realign the spine, and potentially provide spinal cord decompression in a short time. Even today, despite all technological advances, it should not be forgotten and should still be

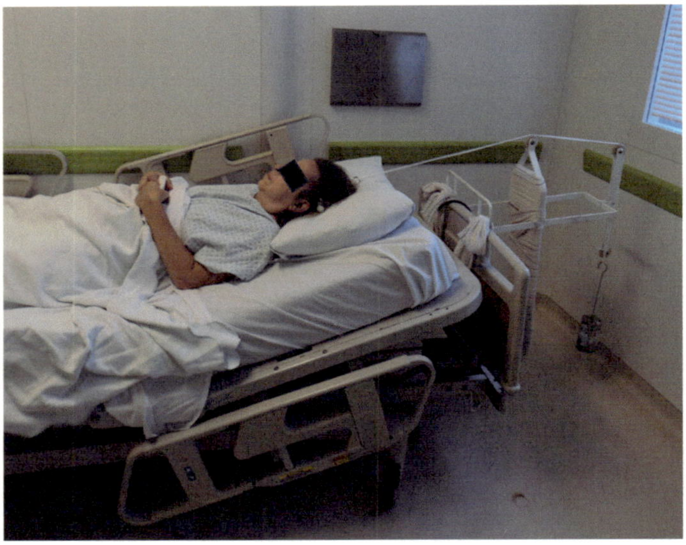

Figure 2.
Closed reduction technique. Note an awake, cooperative, and stable patient. A cranial halo, positioned with pins above the ear, in the projection of the external auditory meatus, approximately 2 cm posteriorly, weighing 1 kg, was used in an attempt to exert a distraction and flexion force vector to unlock the facets.

part of the spine surgeon's armamentarium to remove the patient from neurological emergency and allow surgery and anesthesia to be performed more safely. In addition, performing surgery with an already realigned spine favors shorter operative time and less intraoperative effort. Even the choice of surgical approach (anterior or posterior) may be better defined if reduction has already been achieved. The literature presents high success rates with closed reduction, ranging from 20% to 50% [4, 5]. Not all patients with cervical trauma and facet dislocation are eligible for closed reduction due to the risk of neurological worsening during traction. It is essential that the patient is alert and conscious so that any sign of neurological deterioration leads to immediate suspension and reversal of the procedure. Thus, candidates for closed reduction are:

- Awake, alert, and responsive patient
- Hemodynamically stable
- No evidence of cervical traumatic disc herniation

Contraindications:

- Drowsy patient, decreased level of consciousness, or coma
- Sedated patients
- Drug/alcohol intoxication with decreased consciousness
- Traumatic disc herniation (especially large herniations)

- Findings of other traumatic injuries that may prevent/block reduction

Complications:

- Neurological worsening
- Failure of reduction or incomplete reduction
- Distraction injuries with excessive traction load

6. Operative management

Cervical trauma with facet dislocation is considered an unstable injury due to the failure of at least the facet joint capsule. It is usually associated with spondylolisthesis at that level. Other injuries associated with trauma must be considered for optimal treatment planning. Regarding surgical technique, facet unlocking with vertebral realignment, spinal cord decompression, and arthrodesis must be considered. There is significant controversy in the literature regarding the best way to achieve these objectives. Some authors advocate the anterior approach, others the posterior approach, others the circumferential (combined) approach, and some even advocate a triple approach (anterior–posterior–anterior) in cases with associated traumatic disc herniation. Each approach presents advantages and disadvantages that must be considered in choosing the technique for a given patient, and it is not possible to establish a universal guideline for surgical treatment.

6.1 Anterior cervical approach

The anterior approach presents advantages such as less exposure, less bleeding under usual conditions, lower infection rates, and, for injuries without associated fractures, instrumentation at a single level is relatively simple. However, facet dislocations that have not yet been reduced – either by traction or intraoperative positioning – present greater difficulty for reduction through the anterior approach, requiring more complex maneuvers in an unstable patient. Despite this, reduction is possible in most cases. Discectomy is generally included in this approach, avoiding the risk of spinal cord compression due to traumatic disc herniation during intraoperative reduction maneuvers. Failure of reduction intraoperatively via the anterior approach leads to the need to abort the procedure before fixation, perform a posterior approach for easier facet unlocking, and then return anteriorly to complete fixation (triple approach).

6.2 Posterior cervical approach

The posterior approach presents a major advantage over the anterior approach in that it is easier to correct facet dislocation and, even in cases of failure, allows removal of the facet to enable realignment. This approach also allows associated decompression with laminectomies and facilitates longer arthrodeses. Disadvantages include, in cases of traumatic disc herniation, the risk that realignment maneuvers may cause anterior disc material to compress the spinal cord without direct surgical control. Additionally, it requires wider exposure, with more bleeding, higher infection rates, and longer recovery and rehabilitation times.

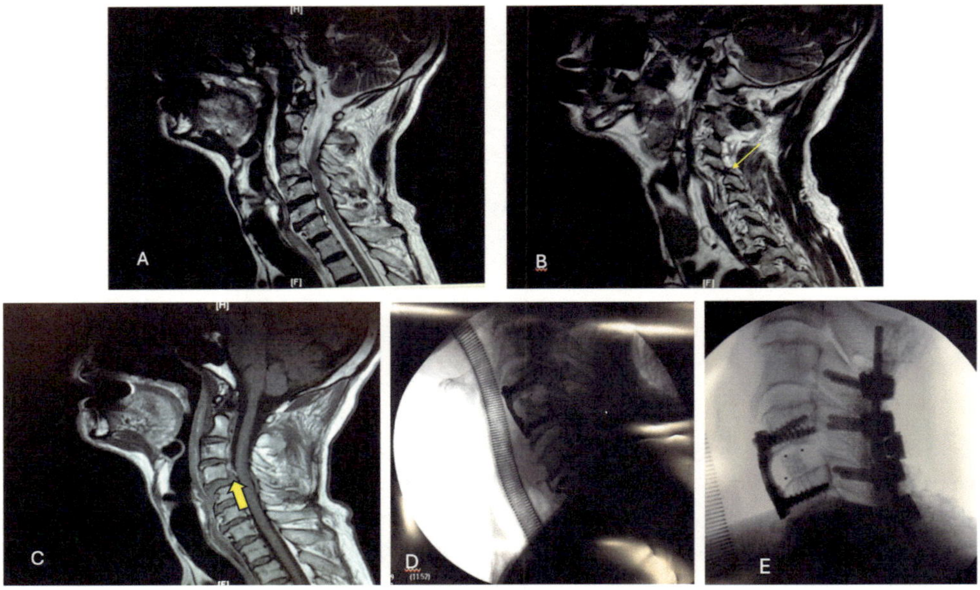

Figure 3.
(A, B, C) sagittal MRI view; (A) traumatic spondylolisthesis C4/5 ; (B) facet dislocation and locking (thin arrow: he articular surfaces of C4 and C5 have lost contact with each other); (C) large traumatic disc herniation C4/5 (thick arrow); (D) intraoperative radioscopy showing good facet reduction with an anterior approach after discectomy C4/C5; (E) posterior fusion (Posterior fusion with lateral mass screws from C3 to C6 to increase fusion rate in combined anterior and posterior arthrodesis).

6.3 Combined (360°) approach

More frequently indicated in cases of facet dislocation associated with other injuries (fractures, ligament injuries). There is discussion as to whether to start with the anterior approach (to remove possible disc herniations, but with greater difficulty in reduction) or with the posterior approach (easier reduction, but higher neurological risk if associated with traumatic disc herniations) as shown in **Figure 3**.

6.4 Triple approach (anterior–posterior–anterior)

Advocated by authors who prefer to initially perform an anterior discectomy, followed by a posterior stage to unlock the facets, reduce and fixate, and a third anterior stage to complete anterior fixation. Systematic reviews attempt to provide evidence regarding the superiority of one approach over another, as well as the risks of posterior approaches, even in the presence of traumatic disc herniations, and the risks of failure of reduction with an isolated anterior approach, which will be addressed in the "Discussion" section.

7. Traumatic disc herniation

The presence of traumatic disc herniation is always a concern in cases of facet dislocation. In theory, reduction maneuvers that attempt to reposition the displaced cranial vertebra could promote mechanical compression of the herniated disc against the spinal cord. In addition, closed reduction using halo traction may also lead to this complication, requiring the patient to be awake, alert, and conscious to report any sensation of neurological worsening during the maneuver. Discussions regarding the surgical approach also take into account the presence of traumatic disc herniation. For this reason, the literature discusses the increased relevance of performing MRI during the investigation. Evidence will be addressed in the "Discussion" section.

8. Timing of intervention

There is extensive discussion in the literature regarding the effect of the timing of surgery after trauma and its prognostic correlation with neurological outcomes [6]. Despite difficulties in obtaining more robust evidence, the literature tends to recommend earlier approaches, preferably within 24 hours after trauma [6–8].

9. Discussion

Cervical fractures with facet dislocation have always been challenging for spine surgeons, both due to neurological risks and the difficulty of reduction and alignment. For reduction, both open and closed techniques have been classically described. The first description of closed reduction was performed by Walton in 1893. The use of halo traction was described by Crutchfield in 1933, and Evans and Kleyn described traction assisted by anesthesia [9]. Theodotou, in the past, recommended reduction using the Gardner-Wells halo only in cases excluding disc herniations. Eismont described the presence of disc herniations in cases of facet dislocation using MRI. Feng advocates performing MRI to investigate traumatic disc herniation in cases of failed closed reduction [10]. Many authors advocate the anterior approach in cases of associated traumatic disc herniation [11]. Hardley et al. published a study of 66 patients with a 58% success rate in realignment with closed reduction. Neurological worsening occurred in 7 patients who underwent open reduction and fixation. There was no neurological worsening in patients who underwent closed reduction while awake [9]. Berrington et al. described 4 cases of neurological worsening after closed reduction in facet dislocations [12]. Eranki et al. observed that patients whose reduction was performed within 4 hours had better neurological recovery than those with reduction after more than 4 hours [13]. Rizzolo et al. described, in a series of 55 patients, a prevalence of 42% of traumatic disc herniation and no neurological worsening after closed reduction [14]. Vaccaro et al. published a series of 11 conscious patients with traumatic disc herniation who underwent closed reduction followed by cervical MRI, with no spinal cord or nerve root compression after traction, relating this to realignment and restoration of the vertebral canal diameter after reduction [15]. In a recent systematic review, Botelho et al. studied 17 papers related to cervical facet dislocation. They described a proportion of traumatic cervical disc herniation of 35% (95% CI = 30 to 41%). The success rate of reduction with cranial traction was approximately 92% (72–98%). The average neurological improvement was 52.66%, described in 15 of 17 articles, estimated as 53% with traction and anterior fixation (AF), 49% with anterior open

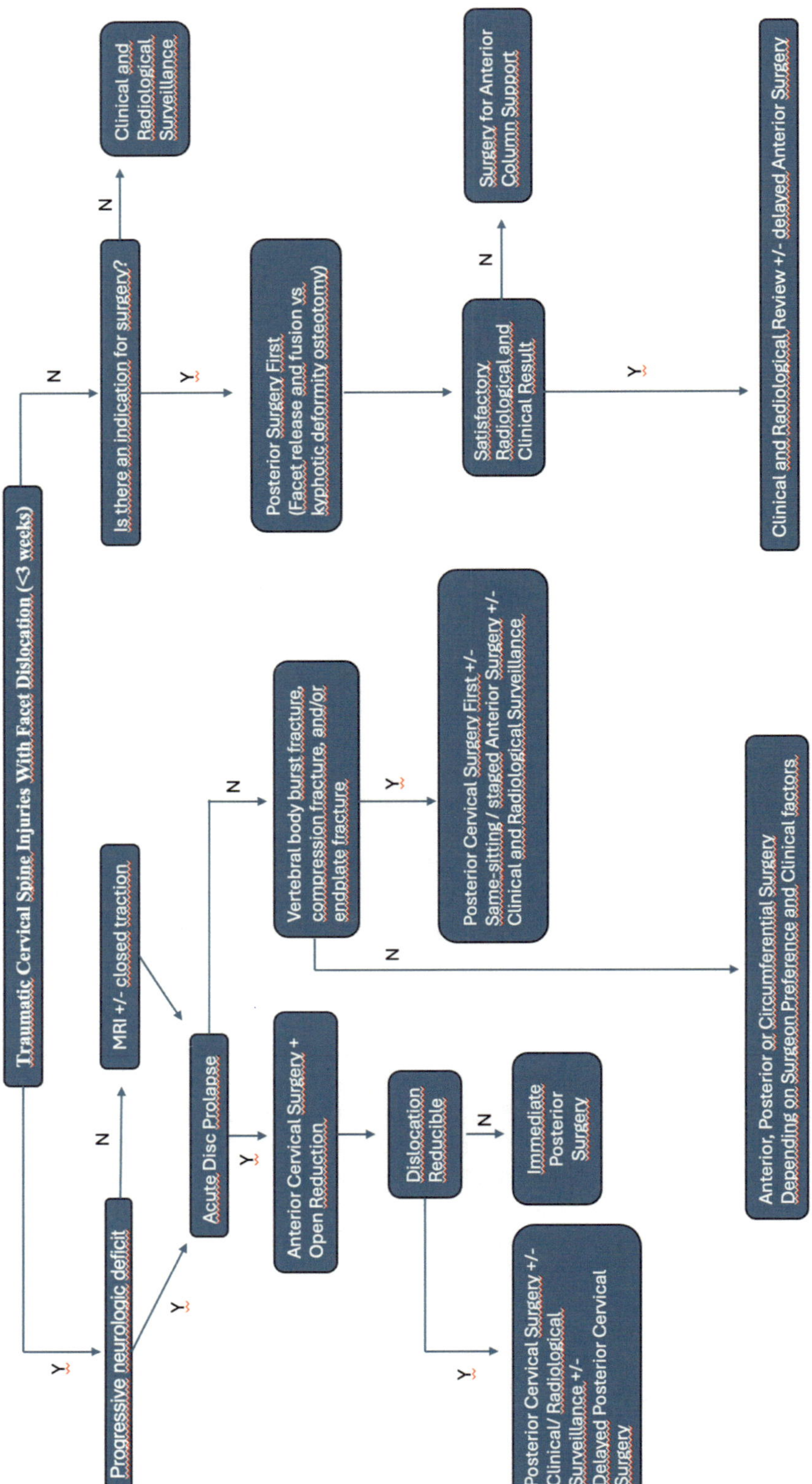

Figure 4.
Evidence-based algorithm proposed by the author, modified from Haq et al., for cervical traumatic facet dislocation (note: Acute fractures (< 3 weeks) have a greater potential for reduction and are considered the primary factor in better response to reduction maneuvers; neurological deficit associated with traumatic disc herniation implies the need for an anterior approach).

reduction (AOR), 58% with posterior fixation (PF), and 45% with posterior open reduction (POR). Neurological worsening was a rare event, with only 7 cases among 679 patients (1%), regardless of the surgical technique chosen [16]. In a systematic review with meta-analysis published by Onishi et al. [17,18], patients with traumatic cervical facet dislocation and traumatic disc herniation treated posteriorly, only 1 case (0.5%) presented neurological worsening, concluding that disc herniations associated with facet dislocation cannot be considered an absolute contraindication for cervical traction or posterior approach for reduction. Liu and Zhang [21] described 3 cases of neurological worsening in 63 patients undergoing AOR. Eranki et al. described 3 of 51 cases undergoing AF, and Newton et al. described 1 patient undergoing traction and PF. Complications occurred in 16 of 608 patients, without the need for reoperations: 5 of 247 in the AOR group (2%) and 11 of 271 in the AF group (4%). The STASCIS trial [7], a prospective cohort study of 313 patients from 6 centers in the United States, compared early (<24 h) versus late surgery in cervical trauma patients and demonstrated objective neurological improvement in patients undergoing early surgery (at least 2 AIS grades improvement at 6 months follow-up). Anderson et al. evaluated variables impacting neurological outcomes in 45 patients with facet dislocation and concluded that reduction timing alone was not significant [19]. AO Spine/Praxis guidelines for the management of cervical spine trauma suggest early surgery (<24 h) and adequate hemodynamic conditions (mean arterial pressure maintained between 75–80 mmHg and 90–95 mmHg for 3–7 days postoperatively), especially in polytrauma patients. Haq et al [20] proposed an evidence-based treatment algorithm. The author of this chapter suggests a modification based on Haq et al., presented in **Figure 4**. Some considerations must be emphasized for the application of the algorithm, particularly related to associated injuries such as fractures, anterior or posterior ligament injuries, ankylosing spondylitis, deformities, segmental kyphosis, spinal cord edema requiring laminectomy and/or duroplasty, and comorbidities (sleep apnea, thyroid tumors, prior anterior or posterior surgeries, tracheostomy). In a Cochrane review, Del Curto et al. analyzed anterior versus posterior approaches in bilateral facet dislocation with a low level of evidence, suggesting better alignment correction with anterior surgery, but no statistical difference in neurological improvement, long-term pain control, or pseudoarthrosis [18]. Botelho et al., in a meta-analysis of surgical treatment for facet dislocation, also found no statistical difference between anterior and posterior approaches [16].

Take-home message

1. MRI plays a crucial role, especially in obtunded patients, in evaluating traumatic disc herniation and predicting the need for anterior decompression.

2. Closed reduction remains valuable, but only in awake, examinable patients.

3. Anterior approaches are favored when traumatic disc herniation is present.

4. Posterior approaches are favored for facet reduction, decompression, and in cases of failed anterior reduction.

5. Combined approaches are valuable in globally unstable or irreducible injuries.

6. Early decompression (<24 h) improves outcomes, aligning with modern principles in spinal cord injury management.

10. Conclusions

Cervical trauma with facet dislocation should be considered complex injuries with a high risk of neurological sequelae, not only at the moment of trauma but also during reduction maneuvers. The presence of traumatic disc herniation, although it raises concern, is not consistently associated in the literature with worse outcomes but must always be evaluated and considered in treatment decision-making. Advances in imaging and surgical techniques, together with growing evidence in the literature, assist in increasingly accurate decision-making, reducing neurological risks and promoting faster recovery.

Author details

Eduardo Bertolini
Department of Neurosurgery, Hospital Do Servidor Público Estadual de São Paulo, São Paulo, Brazil

*Address all correspondence to: edubertolini75@gmail.com

References

[1] Hoffman JR, Wolfson AB, Todd K, Mower WR. Selective cervical spine radiography in blunt trauma: Methodology of the National Emergency X-Radiography Utilization Study (NEXUS). Annals of Emergency Medicine. 1998;**32**:461–469. DOI: 10.1016/s0196-0644(98)70176-3

[2] Stiell IG, Wells GA, Vandemheen KL, et al. The Canadian C-spine rule for radiography in alert and stable trauma patients. Jama. 2001;**286**:1841–1848

[3] Patel MB, Humble SS, Cullinane DC, et al. Cervical spine collar clearance in the obtunded adult blunt trauma patient: A systematic review and practice management guideline from the Eastern Association for the Surgery of Trauma. Journal of Trauma and Acute Care Surgery. 2015;**78**:430–441

[4] Srivastava SK, Aggarwal RA, Bhosale SK, Nemade PS. Neglected dislocation in sub-axial cervical spine: Case series and a suggested treatment protocol. Journal of Craniovertebral Junction and Spine. 2016;7:140–145. DOI: 10.4103/0974-8237.188414

[5] Basu S, Malik FH, Ghosh JD, Tikoo A. Delayed presentation of cervical facet dislocations. Journal of Orthopaedic Surgery. 2011;**19**:331–335. DOI: 10.1177/230949901101900314

[6] Fehlings MG, Tetreault LA, Hachem L, et al. An update of a clinical practice guideline for the management of patients with acute spinal cord injury: Recommendations on the role and timing of decompressive surgery. Global Spine Journal. 2024;**14**:174S–186S DOI: 10.1177/21925682231181883

[7] Fehlings MG, Vaccaro A, Wilson JR, et al. Early versus delayed decompression for traumatic cervical spinal cord injury: Results of the Surgical Timing in Acute Spinal Cord Injury Study (STASCIS). PLoS ONE. 2012;**7**:e32037

[8] Kwon BK, Tetreault LA, Martin AR, et al. A clinical practice guideline for the management of patients with acute spinal cord injury: Recommendations on hemodynamic management. Global Spine Journal. 2024;**14**:187S–211S

[9] Hadley MN, Walters BC, Grabb BC, Oyesiku NM, Przybylski GJ, Resnick DK, et al. Initial closed reduction of cervical spine fracture-dislocation injuries. Neurosurgery. 2002;**50**(3 Suppl):S44–50. DOI: 10.1097/00006123-200203001-00010

[10] Feng G, Hong Y, Li L, Liu H, Pei F, Song Y, et al. Anterior decompression and nonstructural bone grafting and posterior fixation for cervical facet dislocation with traumatic disc herniation. Spine. 2012;**37**(25):2082–2088. DOI: 10.1097/BRS.0b013e31825ee846

[11] Eismont FJ, Arena MJ, Green BA. Extrusion of na intervertebral disc associated with traumatic subluxation or dislocation of cervical facets. Case Report. Journal of Bone and Joint Surgery. American Volume. 1991;**73**(10):1555–1560. DOI: 10.2106/00004623-199173100-00014

[12] Berrington NR, van Staden JF, Willers JG, van der Westhuizen J. Cervical intervertebral disc prolapse associated with traumatic facet dislocations. Surgical Neurology. 1993;**40**(5):395–399. DOI: 10.1016/0090-3019(93)90219-q

[13] Eranki V, Koul K, Mendz G, Dillon D. Traumatic facet jointdislocation in Western Australia. European Spine Journal. 2016;**25**(4):1109–1116. DOI: 10.1007/s00586-014-3627-x

[14] Rizzolo SJ, Piazza MR, Cotler JM, Balderston RA, Schaefer D, Flanders A. Intervertebral disc injury complicating cervical spine trauma. Spine. 1991;**16**(6 Suppl):S187–S189. DOI: 10.1097/00007632-199106001-00002

[15] Vaccaro AR, Falatyn SP, Flanders AE, Balderston RA, Northrup BE, Cotler JM. Magnetic resonance evaluation of the intervertebral disc, spinal ligaments, and spinal cord before and after closed traction reduction of cervical spine dislocations. Spine. 1999;**24**(12):1210–1217. DOI: 10.1097/00007632-199906150-00007

[16] Botelho RV, de Freitas Bertolini E, Barcelos ACES, et al. The surgical treatment of subaxial acute cervical spine facet dislocations in adults: A systematic review and meta-analysis. Neurosurgical Review. 2022;**45**:2659–2669

[17] Onishi FJ, Daniel JW, Joaquim AF, et al. The impact of traumatic herniated discs in cervical facets dislocations treatments: Systematic review and meta-analysis. European Spine Journal. 2022;**31**:2664–2674

[18] Del Curto D, Tamaoki MJ, Martins DE, et al. Surgical approaches for cervical spine facet dislocations in adults. The Cochrane Database of Systematic Reviews. 2014;**2014**: CD008129

[19] Greg Anderson D, Voets C, Ropiak R, et al. Analysis of patient variables affecting neurologic outcome after traumatic cervical facet dislocation. The Spine Journal. 2004;**4**:506–512

[20] Haq IZ, Fehlings MG, Vaccaro AR, Furtado NR, Wang B, Upadhyayula P, LaFuente J, Casey ATC, Allibone JB, Ganau M. Traumatic bilateral cervical facet dislocation:a comprehensive review and evidence-based treatment algorithm. European Spine Journal. 2025. DOI: 10.1007/s00586-025-09503-7

[21] Liu K Zhang Z Comparison of a novel anterior-only approach and the conventional posterior–anterior approach for cervical facet dislocation: A retrospective study *European Spine Journal*. 2019;**28**:2380–2389. DOI: 10.1007/s00586-019-06073-3

Section 2

Minimally Invasive, Endovascular, and Functional Neurosurgery

Chapter 6

Beyond the Burr-Hole: A Tailored Approach to Chronic Subdural Haematoma Treatment Using Middle Meningeal Artery Embolisation

Yevgeny Karepov, Youssef Chedid and Conor Gillespie

Abstract

Chronic subdural haematoma (cSDH) is amongst the most common neurosurgical conditions, with recurrence rates after burr-hole evacuation ranging 10–30%. Middle meningeal artery embolisation (MMAE) has emerged as a promising pathophysiologically targeted therapy designed to interrupt the vascular drivers of recurrence. This chapter reviews the historical evolution, underlying mechanisms, technical considerations, and clinical evidence surrounding MMAE, integrating data from randomised controlled trials, large observational series, and meta-analyses. Predictors of success and failure, complication profiles, and the health-economic implications of adopting MMAE in neurosurgical practice are critically appraised. Future research directions, including ongoing multi-centre trials and technological innovations, are also discussed. The chapter aims to provide an authoritative, evidence-based synthesis for neurosurgeons evaluating the role of MMAE in modern cSDH management in order to tailor the best treatment for every patient.

Keywords: cSDH, MMAE, surgery, adjunct, chronic subdural haematoma, middle meningeal artery embolisation, endovascular neurosurgery, health economics

1. Introduction

Chronic subdural haematoma (cSDH) is the most frequently encountered neurosurgical condition nowadays, particularly in the elderly. Its incidence continues to rise as a result of global population ageing and the widespread use of antithrombotic and anticoagulant therapies [1, 2]. Despite the relative simplicity of burr-hole drainage and related techniques, recurrence rates remain significant, reported between 10% and 30%, and recurrent cases are especially challenging in frail or medically complex patients [3, 4]. These factors combine to make cSDH a condition of growing clinical and societal importance worldwide.

In order to address this backdrop, middle meningeal artery embolisation (MMAE) has emerged as a promising therapeutic strategy, either as an adjunct to surgical evacuation or, in selected cases, as a standalone primary intervention. The rationale derives from the pathophysiology of cSDH: fragile neovascular membranes, mostly supplied by the MMA, are thought to sustain the haematoma through ongoing cycles of microhaemorrhage and inflammation [5]. Embolisation targets this pathological vascular supply, with the intent of breaking this cycle and, as a result, accelerating haematoma resolution and reducing the risk of recurrence.

Recent clinical evidence supports the potential of MMAE, though findings vary by patient population. Early randomised trials and pooled analyses indicate that MMAE is safe and may reduce haematoma progression or recurrence, with signals of greater benefit in non-surgical cohorts compared with patients undergoing surgery, in whom trial results have been less consistent [6–8]. Large observational series further demonstrate radiographic improvement in over 85% of cases, with low clinical failure rates and procedure-related complication rates generally below 5% [9, 10]. Collectively, these data position MMAE as a safe and valuable addition to the therapeutic armamentarium, particularly in recurrent cSDH and in patients at elevated surgical risk, while its definitive role alongside conventional surgery continues to be clarified through ongoing trials.

2. Historical background

The clinical understanding and management of cSDH have evolved substantially over centuries. In the eighteenth century, Scottish surgeon James Hill (1703–1776) provided an early, systematic description of subdural and epidural haematomas, recognising lucid intervals and the need for trepanation with drainage to alleviate cerebral compression [11]. Hill's work laid critical groundwork for appreciating the pathophysiology of subdural collections and the role of timely surgical decompression.

For much of the twentieth century, therapeutic approaches remained firmly surgical: simple burr-hole drainage or craniotomy with membrane removal, often under general anaesthesia. Although technically straightforward, these methods were hampered by high recurrence rates, sometimes approaching 30%, particularly in elderly or frail patients on anticoagulants [3]. In parallel, advances in interventional neuroradiology, pioneered by figures such as Ayub Ommaya, who helped develop percutaneous spinal angiography and embolisation techniques for arteriovenous malformations, set the stage for endovascular treatment entering neurosurgery [12].

The true origin of MMAE for cSDH is relatively recent. The first reported case in the peer-reviewed literature emerged in 2000, when Mandai et al. described a successful embolisation in a patient with refractory cSDH, marking the inaugural clinical application of MMAE as an alternative or adjunctive strategy to surgery [13]. Combining all these data, Tanaka and colleagues provided histopathological evidence suggesting connections between the outer neovascular membrane vessels of cSDH and the MMA, offering an anatomical rationale for targeting these vessels via embolisation [5].

Over the subsequent two decades, growing case series and comparative studies, including pioneering work by Ban et al. and Link et al., demonstrated efficacy in reducing recurrence, promoting haematoma resolution, and offering a minimally invasive option for high-risk surgical patients [9, 14]. These collective developments propelled MMAE from a novel idea to a validated intervention in contemporary neurosurgical practice.

2.1 Trials review – search strategy and selection criteria

In order to optimise the data for this chapter, we used our team database, which was published in recently published meta-analyses by Gillespie et al. [7]. To obtain the data, we conducted a systematic review of ongoing trials according to the Preferred Reporting Items for Systematic Reviews and Meta-Analyses (PRISMA) guidelines (PROSPERO [CRID TBC], [**Figure 1**]). We searched 19 trial databases. Population, Intervention, Comparator, Outcome, Study Design (PICOS) criteria were used (**Table 1**) while searching the databases. The trial categories, estimated population size, and completion dates are shown in **Figure 2** as a bubble plot. Quality assessment was completed for recently published three major RCT, as part of the meta-analysis. Quality assessment was done using the Cochrane risk of bias (RoB2) tool. The quality assessment showed low risk for randomisation and selection of the reported results, as well as some concerns for interventions and outcomes due to non-blinding. Quality assessment was not completed for the rest of the study, as the aim was descriptive manner.

3. Pathophysiology of cSDH

The pathophysiology of cSDH is now recognised as a complex, dynamic process involving vascular, inflammatory, and fibrinolytic mechanisms, rather than a static accumulation of liquefied clot. Traditionally, cSDH was thought to arise from traumatic tearing of bridging veins, with subsequent gradual expansion due to the persistence of liquefied blood products in the subdural space [15]. However, accumulating histopathological and radiological evidence has shifted this paradigm toward a model centred on a pathological vascularised neomembrane that perpetuates haematoma growth.

The subdural collection is lined by an outer and inner membrane. The outer membrane is biologically active, containing fragile neocapillaries derived primarily from the MMA and characterised by incomplete basement membranes and leaky endothelial junctions [5]. Fragile vessels are prone to microhaemorrhages, contributing to the cyclical expansion of the haematoma. Inflammatory mediators, including vascular endothelial growth factor (VEGF), interleukins, and matrix metalloproteinases, further drive angiogenesis, vascular permeability, and extracellular matrix remodelling [16, 17].

Simultaneously, impaired coagulation and persistent local fibrinolysis create an environment leading to bleeding and resorption imbalance, favours the expansion of haematoma (**Figure 3**). Elevated levels of tissue plasminogen activator and fibrin degradation products have been identified within haematoma fluid, reflecting ongoing fibrinolysis and impaired clot stability [18]. This pro-haemorrhagic environment explains the limited efficacy of simple haematoma drainage without addressing the vascular source, leading to a tendency for recurrence.

Radiological and histological studies confirm that the vascular supply of the outer membrane is predominantly derived from the MMA, providing a clear mechanistic rationale for targeting this artery in treatment [19]. By occluding these pathological

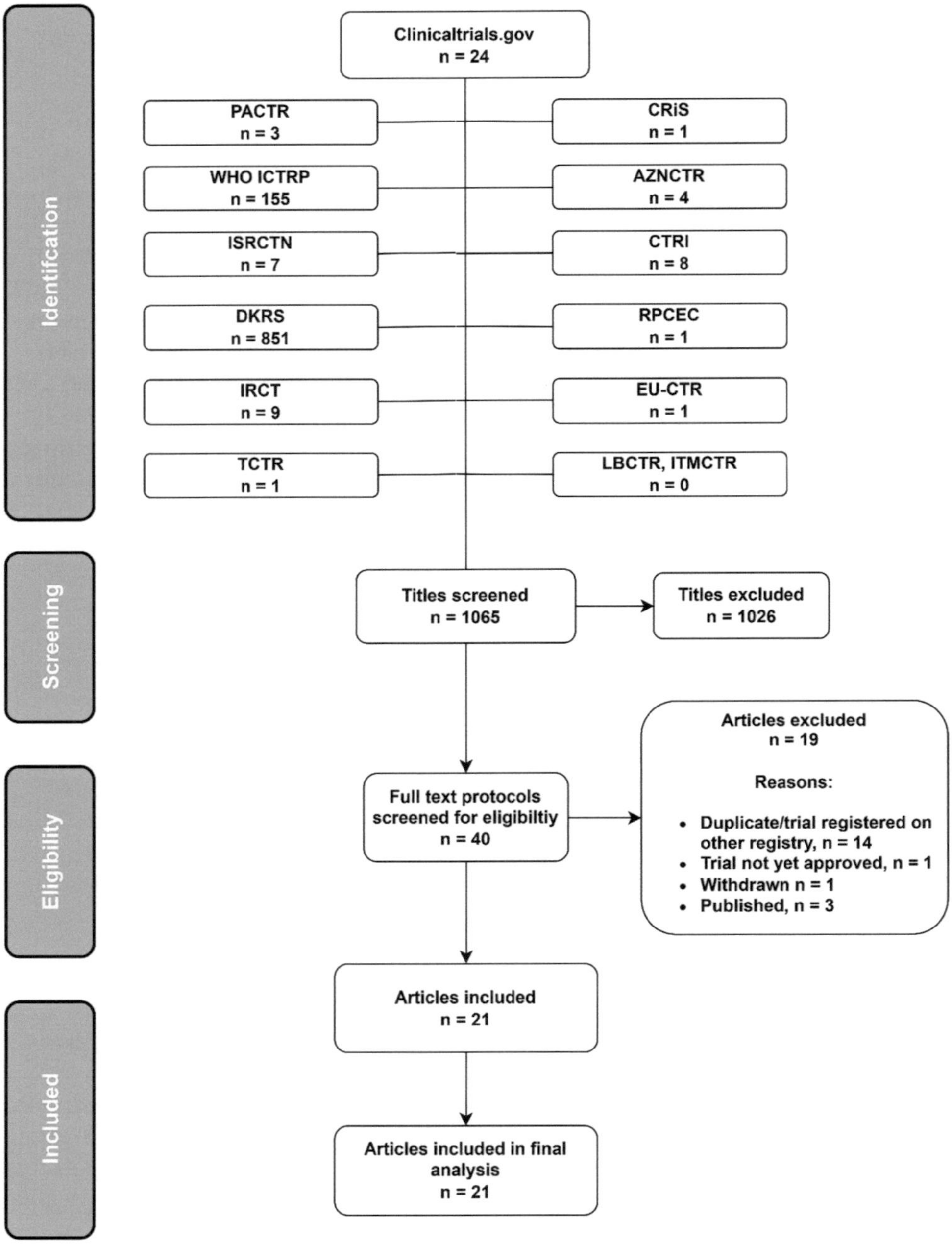

Figure 1.
PRISMA flow-chart.

feeders, MMA embolisation interrupts the cycle of microhaemorrhage and inflammation, thereby promoting haematoma resolution and reducing recurrence risk. This mechanistic foundation underpins the growing clinical interest in MMAE as a disease-modifying therapy rather than merely a decompressive drainage procedure.

Review question	**What are the current ongoing trials investigating middle meningeal artery embolization for chronic subdural hematoma, and what are the main outcome measures?**
Population	Adults ≥18 years with CSDH
intervention	Either: • MMAE alone • Surgery + MMAE
Comparator	Either: • Surgery alone • Conservative management/best medical care
Outcomes	Primary: • Identify the primary outcome used in trials • Identify definitions used in ongoing trials Secondary: • Identify the number of ongoing trials • Identify methodologies (including trial inclusion and exclusion criteria)
Setting	Studies taking place in any neurosurgery, trauma, or radiology/neurology department or center.
Study design	Must be a randomized trial, with an intervention and a control group – exclude prospective cohort studies or studies with a single arm. Include studies regardless of blinding, as long as there is random allocation to at least two groups.
Follow-up	Any follow-up

Table 1.
Population, intervention, comparator, outcome, study design (PICOS) criteria used for the review by Gillespie et al. [7].

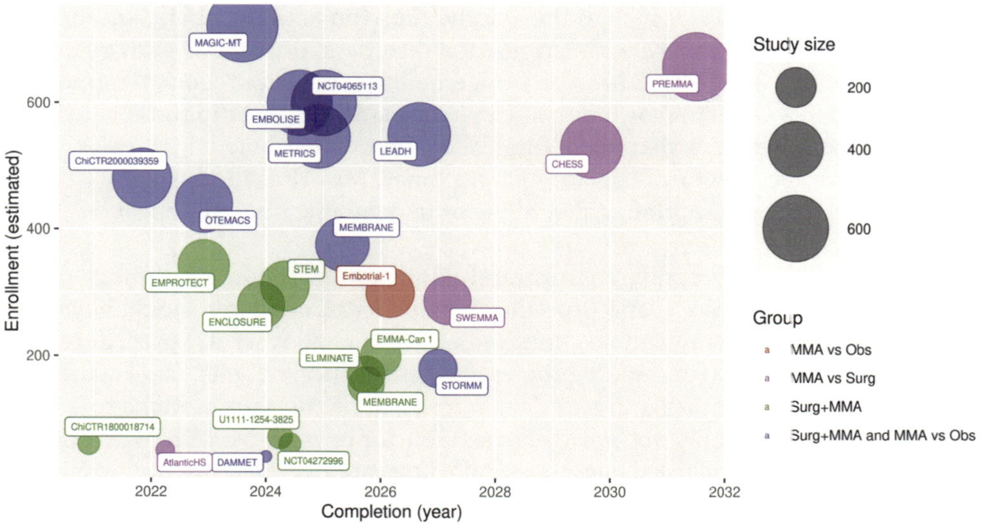

Figure 2.
*Bubble plot of ongoing and recently published MMAE trials, grouped by trial design/arms. Bubble size = reflective of the total population size of the trial. *MMA vs. obs = trial of MMAE vs. nonsurgical management only; MMA vs. surg = trial of MMAE alone vs. surgery; surg + MMA = trial of MMAE + surgery vs. surgery; surg + MMA, and MMA vs. obs = trial including both arms.*

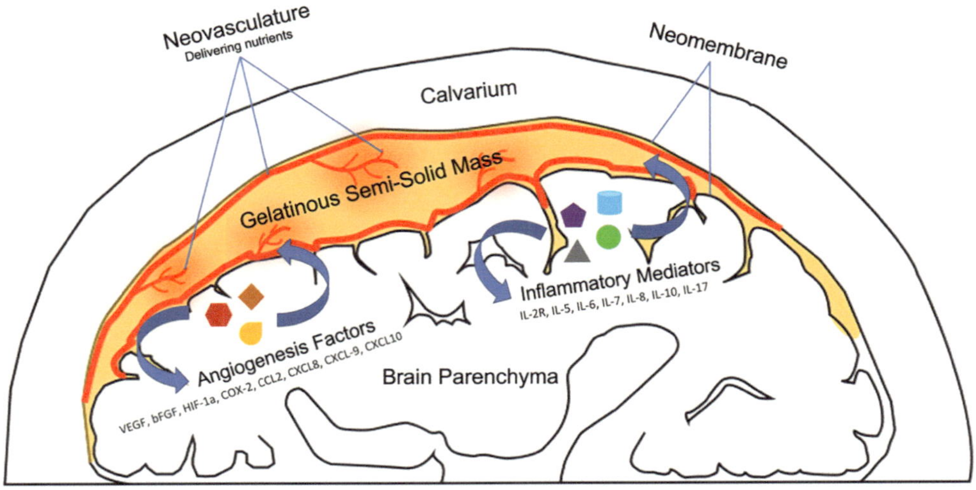

Figure 3.
cSDH expansion occurs via both haemodynamic elements of haemorrhage, and biochemical factors involving cytokine signalling cascades and angiogenesis that promote growth in a mass-like manner (as presented in Interventional Neuroradiology, 1–6, 2024 [20].

4. Indications and patient selection

The clinical role of MMAE in cSDH is still being refined, but several indications have emerged from current evidence and practice. The most widely accepted use is as an adjunct to burr-hole drainage to reduce recurrence risk, particularly in patients at elevated risk of relapse, such as the elderly, those on antithrombotic and/or anticoagulant therapy, those with coagulation disorders, or patients with bilateral haematomas [21, 22]. MMAE has also been employed as a stand-alone treatment in selected patients with mild or minimally symptomatic cSDH, or in those with significant comorbidities that make surgical evacuation extremely high risk [23]. In the context of recurrent or refractory haematomas, MMAE is increasingly considered a preferred option, either alone or in combination with repeat evacuation [24].

Patient selection typically requires careful integration of clinical status, haematoma characteristics, and procedural feasibility. Candidates should have accessible MMA anatomy and no contraindications to endovascular procedures. Conversely, patients with rapidly progressive neurological deficits, large acute-on-chronic collections requiring urgent decompression, or contraindications to angiography are generally not considered suitable for primary MMAE. With ongoing randomised trials, formalised consensus guidelines are expected to provide more standardised criteria in the forthcoming exciting era of cSDH treatment.

5. Technical considerations

MMAE is most commonly accessed via trans-femoral or trans-radial arterial access, with a guiding catheter advanced into the external carotid artery, then to the maxillary artery and subsequent super-selective catheterisation of the MMA using a microcatheter–microwire system [25]. Preparation using computed tomography angiography (CTA) of the head and neck and biplane digital subtraction angiography (DSA) is essential to delineate arterial anatomy, evaluate for variant branches, and exclude hazardous anastomoses with the ophthalmic artery, internal carotid artery, or posterior circulation. Particular attention is required to avoid inadvertent embolisation of orbital or petrosal branches, which may result in visual or cranial nerve deficits if not recognised [9].

A variety of embolic agents have been employed, reflecting differences in operator experience and institutional practice. Polyvinyl alcohol (PVA) particles and calibrated microspheres in the 150–300 μm range remain the most widely used, offering distal penetration with a favourable safety profile [14]. Liquid embolics, such as n-butyl cyanoacrylate or ethylene-vinyl alcohol copolymer, allow more durable occlusion but require technical expertise to mitigate the risks of reflux or cranial nerve ischemia [26]. Proximal coil occlusion has also been described, though it is rarely sufficient in isolation due to persistent collateral supply [27]. Regardless of the material, durable embolisation of distal MMA branches supplying the outer membrane of the haematoma appears crucial to reducing recurrence.

Patients are usually managed under conscious sedation or general anaesthesia depending on comorbidity and cooperation. Peri-procedural systemic heparinisation is frequently administered, though antiplatelet therapy is not routinely required unless adjunctive stent-based techniques are used [21]. Post-procedure monitoring typically includes a short inpatient stay, with follow-up imaging within several weeks to assess haematoma resolution and detect rare complications such as ischemic events or procedure-related haemorrhage. Collectively, these technical refinements have enabled MMAE to be performed safely and reproducibly in a broad range of patients.

6. Clinical outcomes

The clinical efficacy and safety of MMAE for cSDH have been investigated through randomised controlled trials, meta-analyses, and large observational studies. The first high-quality evidence emerged from randomised trials, including MAGIC-MT, EMBOLISE, and STEM, which together enrolled more than 1,400 patients. Pooled analysis of these trials demonstrated that MMAE did not significantly reduce recurrence in patients treated adjunctively following surgical evacuation, though a significant benefit was observed in non-surgical cohorts, where embolisation reduced symptomatic progression and recurrence without adversely affecting functional outcomes (**Figure 4**) [6–8].

Subsequent systematic reviews and meta-analyses integrating randomised and observational data have generally supported these findings. MMAE combined with usual care was associated with a lower risk of haematoma recurrence compared with usual care alone, though no consistent differences were identified in 90-day functional outcomes such as modified Rankin Scale scores [28]. A larger meta-analysis of six randomised trials involving more than 1,500 patients similarly demonstrated reduced recurrence and surgical rescue interventions in patients undergoing MMAE, but also reported a modest increase in infection rates, highlighting the need for careful patient monitoring and standardised peri-procedural protocols [29].

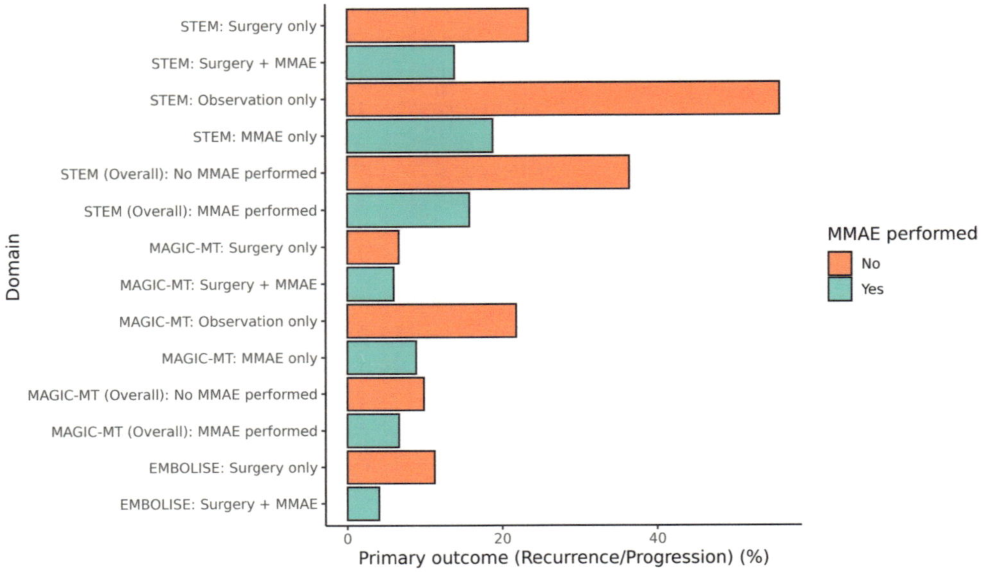

Figure 4.
Chart demonstrating differences in primary outcome MMAE groups (red) compared to non-MMAE groups as controls (blue), stratified by cohort (overall, surgical, and nonsurgical) as presented in Gillespie et al. [7].

Observational studies have provided complementary evidence, particularly in the surgical population. Propensity-matched analyses have shown that adjunctive MMAE following burr-hole evacuation significantly reduces recurrence rates compared with surgery alone, with recurrence typically in the range of 5–8% compared with 20–30% for surgery without embolisation [9]. Registry-based series have further suggested that standalone MMAE in non-surgical patients is associated with reduced long-term mortality compared with conservative management, although the rate of subsequent surgical rescue does not appear to differ significantly between the groups [30].

Taken together, these studies indicate that MMAE is a safe and effective intervention that can reduce haematoma recurrence and re-intervention in selected populations, particularly in patients managed conservatively or at high risk for re-accumulation. Its role as a universal adjunct to surgical evacuation remains less certain, with mixed evidence from randomised trials, but mounting observational data continue to support potential benefit in selected high-risk surgical cohorts.

7. Complications of MMAE

While MMAE for cSDH is generally well-tolerated, it is not devoid of risks. Across large multi-centre cohorts and randomised trials, reported complication rates range between 2% and 5%, most of which are minor and self-limiting. The most frequently observed adverse events include transient headache, groin haematoma, and access-site complications, which occur in fewer than 3% of patients [9, 31].

More serious but rare complications include ischemic stroke, cranial nerve palsy, injury to the femoral artery, and inadvertent embolisation of non-target arteries such as the ophthalmic or petrosal branches. Embolisation material migration into the ophthalmic circulation carries a risk of visual loss, though this has been reported in less than 0.2% of cases [25]. Cranial nerve deficits, typically involving the abducens or facial nerve, have similarly been described but are exceedingly uncommon [27].

Importantly, procedural mortality attributable to MMAE is vanishingly rare. In the landmark randomised controlled trials, including the STEM and EMBOLISE studies, no procedure-related deaths were reported, and serious morbidity directly linked to embolisation remained under 1% [6, 8]. Adverse events do occur, and they are often related to anatomical variants of the middle meningeal artery or inadvertent embolic penetration beyond the intended territory. Meticulous pre-procedural angiographic evaluation and technique are critical [24].

The available evidence suggests that MMAE is a safe intervention, with an overall complication profile favourable to surgical alternatives, particularly in elderly or anticoagulated patients. The emphasis, however, remains on meticulous patient selection and technical precision to minimise rare but potentially severe events.

8. Predictors of MMAE success and failure

Several technical and clinical factors have been identified as predictors of MMAE efficacy. Distal penetration of embolic material into the vascular network supplying the outer haematoma membrane is associated with a lower recurrence rate and higher rates of radiographic resolution. On the other hand, incomplete embolisation or proximal coil occlusion alone correlates with poorer outcomes [14, 25].

Vessel diameter has also been shown to influence efficacy, with MMA calibres <1.5 mm limiting adequate penetration and increasing the risk of treatment failure [32].

Clinical variables, including advanced age, bilateral haematomas, and the use of antithrombotic or anticoagulant therapy, have been independently linked to higher recurrence rates despite embolisation [10]. Patients receiving direct oral anticoagulants have demonstrated approximately double the risk of recurrence compared with non-anticoagulated cohorts [30]. Furthermore, residual haematoma thickness greater than 15 mm on post-procedural imaging is an important predictor of progression and the need for further surgical intervention [26].

These findings highlight the importance of careful patient selection and consideration of clinical comorbidities when planning MMAE, alongside the previously mentioned meticulous angiographic preparation and technique.

9. Health-economic impact of MMAE

cSDH is associated with a significant economic burden due to its high incidence, frequent recurrence, and the need for repeat hospitalisation and surgery. In Europe and North America, the annual cost of cSDH is substantial, particularly in elderly populations, where prolonged inpatient stays and rehabilitation requirements are extremely common [1]. Conventional burr-hole drainage carries up to a 30% recurrence rate, which substantially increases healthcare expenditure through reoperations and extended recovery periods.

Cost-effectiveness analyses indicate that MMAE has the potential to reduce these costs. In a US model, adjunctive MMAE decreased the requirement for repeat surgery by nearly half, leading to cost savings despite the upfront expense of embolisation [33]. A European evaluation found MMAE to be cost-neutral within the first year of treatment owing to shorter hospital stays and lower recurrence rates [22].

The greatest economic benefit appears in elderly or medically fragile patients, where the risks and costs of recurrent surgery are highest. These findings suggest that, in addition to improving clinical outcomes, MMAE may represent a value-based strategy for resource allocation in neurosurgical practice.

10. Future directions

Research into MMAE for cSDH is progressing rapidly, with several areas likely to shape its future role in neurosurgical practice. Large randomised controlled trials currently underway across North America, Europe, and Asia will be pivotal in establishing definitive evidence regarding the efficacy of MMAE as both an adjunct to surgery and a stand-alone therapy [33]. These studies are expected to refine patient selection criteria and inform clinical guidelines.

Technological advances are also anticipated to enhance procedural safety and efficacy. Innovations in microcatheter systems, embolic agents, and high-resolution imaging modalities, including cone-beam CT and 3D rotational angiography, enable greater precision in identifying pathological vascular networks while minimising the risk of non-target embolisation [34, 35]. As these tools evolve, they may reduce complication rates and broaden the pool of patients suitable for MMAE.

Future strategies may also involve combination therapies. Early explorations of MMAE performed alongside minimally invasive surgical evacuation, targeted anti-angiogenic agents, or modulation of local fibrinolytic pathways suggest the possibility of synergistic benefits in haematoma resolution and recurrence prevention [36]. Parallel to this, biomarker-based approaches for predicting recurrence risk or embolisation response are under investigation, potentially paving the way for personalised treatment algorithms [37].

Long-term outcomes remain an important knowledge gap. Beyond recurrence rates and radiographic resolution, ongoing studies are beginning to evaluate functional independence, cognitive trajectories, and quality-of-life outcomes, as well as health-economic considerations in aging populations [38]. The integration of these data will ultimately determine whether MMAE becomes a first-line treatment option or a highly selective adjunct in the management of cSDH.

11. Conclusion

MMAE represents a giant leap in the treatment of cSDH, offering a pathophysiologically targeted approach that addresses the vascular drivers of recurrence. While burr-hole drainage remains the mainstay for symptomatic or large haematomas, MMAE has emerged as a valuable adjunct in high-risk patients and a promising alternative in select non-surgical cases.

Evidence from randomised trials, meta-analyses, and large observational studies demonstrates that MMAE can reduce recurrence rates and the need for reoperation, with a favourable safety profile and low procedure-related morbidity. Importantly, predictors of treatment success have been defined, including adequate distal embolisation and patient-specific factors such as haematoma size and anticoagulant use, which aid in refining case selection. Increasing health-economic data further suggest that MMAE may not only improve outcomes but also reduce overall healthcare costs by decreasing reoperation rates and shortening hospital stays.

As the global burden of cSDH rises simultaneously with population ageing and widespread antithrombotic use, treatment strategies must adapt to balance efficacy, safety, and resource allocation. MMAE is well-positioned to become a key component of this evolving treatment paradigm. With further standardisation of the technique and confirmation from ongoing multicentre randomised trials, MMAE is likely to transition from an emerging option to an integral part of evidence-based, personalised care for patients with cSDH.

12. Our experience in treating cSDH

Our centre serves a high-volume catchment area (serving as a referral neurosurgical centre for about six million people), managing cases of cSDH daily (285 cases in the last year). Most of the cases admitted to our centre are surgical cases, while most of the non-surgical cases are managed in district general hospitals.

MMAE is still an emerging procedure and has not yet been established as a gold standard in the treatment of cSDH in our centre; thus, the data relating to MMAE and cSDH in our centre has been collected over the past year with ongoing database collection.

Currently, moderate-to-large or symptomatic cSDHs are treated via surgical burr-hole evacuation to release the pressure and achieve cortical decompression (264 cases – 92.6%). All patients have a subdural or sub-galeal drain in place for 24–48 hours (100%).

While we do not routinely perform MMAE as an adjunct procedure – consistent with recent meta-analyses (e.g., Gillespie et al. [7]) showing no statistically significant benefit in unselected cohorts – only 14 had surgical evacuation with adjunct MMAE (5.3%).

We do offer it to high-risk subgroups. This includes elderly patients, those with bilateral haematomas, coagulation disorders, or those requiring rapid resumption of antithrombotic/anticoagulant therapy. (All 14 – 100% of adjunct MMAE belongs to the high-risk subgroup.) Given that patients on direct oral anticoagulants face nearly double the recurrence risk, adjunct MMAE is discussed for those patients.

Currently, our approach is to discuss adjunct MMAE in cases of recurrent cSDH requiring a second surgical intervention (32% of recurrent interventions underwent MMAE as well), or for asymptomatic patients with radiological evidence of recurrence, provided there are no contraindications (20% of radiological recurrences without surgical intervention had MMAE).

Standalone MMAE remains a viable strategy and is discussed for growing on follow-up scans or asymptomatic collections in high-risk patients, particularly those with impaired coagulation. Standalone MMAE was performed as initial treatment for cSDH or for evidence of recurrence without surgical intervention (**Table 2** summarises last year's cSDH treatment in our centre).

Although complications are rare in MMAE and are mainly related to pain and local haematoma at the puncture site (4/33, 12%), two major complications occurred during the past year (6%), including loss of vision and one major vascular event in the punctured artery.

Ultimately, our management of the increasing cSDH burden is highly individualised. A tailor-made approach for the management of cSDH is the current gold standard practice for our patients. While MMAE is generally safe, the potential for rare but serious complications requires a rigorous risk-benefit analysis before proceeding with the tailor-made treatment plan. As clinical data matures, MMAE is becoming a cornerstone of the cSDH treatment algorithm, whether as an adjunct or a standalone intervention for appropriate patients.

13. Practice points

- cSDH recurrence remains common after burr-hole evacuation, reported in up to 30% of cases.
- MMAE reduces recurrence to ~5–10% in most contemporary series and meta-analyses.
- Predictors of recurrence after MMAE include small MMA calibre (<1.5 mm), anticoagulant or antiplatelet use, bilateral haematomas, and residual haematoma thickness >15 mm).
- Major complications are rare (<2%) but may include ischemic stroke, major limb vascular injury, or cranial nerve palsy due to non-target embolisation.
- Cost-effectiveness analyses suggest MMAE reduces overall healthcare expenditure by lowering re-operation rates and hospital stay length, particularly in elderly or high-risk patients.

	Number of patients	Recurrence rate
CSDH treated operatively	264	35 (13.3%)
CSDH treated conservatively	21	0
Surgery + MMAE	14	3 (25%)
Standalone MMAE	11	2 (18.2%)
Recurrent surgery	25	2 (8%)
MMAE after recurrence with surgery	8	1 (12.5%)
MMAE after recurrence without surgical intervention	2	0

Table 2.
Summary of treatment of cSDH, recurrence rate, and MMAE in our centre for the last year.

Acknowledgments

We are profoundly grateful for the mentorship of Professor Peter J.A. Hutchinson, whose insights and expertise were instrumental in the development of this chapter. His visionary leadership and pioneering work in the field of neurosurgery, and especially in neurotrauma, provided a constant source of inspiration and intellectual rigor. We thank him for his guidance and for fostering an environment of academic excellence that made this work possible. We are deeply grateful to all our colleagues, mentors, and staff for their support and collaboration. Their commitment to excellence and willingness to share their expertise were essential to the completion of this chapter.

The author acknowledges the use of the Gemini AI tool for limited language polishing of the manuscript.

Conflict of Interest

The authors declare no conflict of interest.

Author details

Yevgeny Karepov[1*], Youssef Chedid[1,2] and Conor Gillespie[1,2]

1 Department of Neurosurgery, Cambridge University Hospitals NHS Foundation Trust, Cambridge, UK

2 Department of Clinical Neurosciences, University of Cambridge, Cambridge, UK

*Address all correspondence to: karepovmd@gmail.com

References

[1] Rauhala M, Helén P, Huhtala H, Heikkilä P, Iverson GL, Niskakangas T, Öhman J, Luoto TM. Chronic subdural hematoma-incidence, complications, and financial impact. Acta Neurochirurgica. 2020;**162**(9):2033–2043. DOI: 10.1007/s00701-020-04398-3

[2] Almenawer SA, Farrokhyar F, Hong C, Alhazzani W, Manoranjan B, Yarascavitch B, Arjmand P, Baronia B, Reddy K, Murty N, Singh S. Chronic subdural hematoma management: A systematic review and meta-analysis of 34,829 patients. Annals of Surgery. 2014;**259**(3):449–457. DOI: 10.1097/SLA.0000000000000255

[3] Santarius T, Kirkpatrick PJ, Ganesan D, Chia HL, Jalloh I, Smielewski P, Richards HK, Marcus H, Parker RA, Price SJ, Kirollos RW, Pickard JD, Hutchinson PJ. Use of drains versus no drains after burr-hole evacuation of chronic subdural haematoma: A randomised controlled trial. Lancet. 1067-73;**374**(9695). DOI: 10.1016/S0140-6736(09)61115-6

[4] Brennan PM, Kolias AG, Joannides AJ, Shapey J, Marcus HJ, Gregson BA, Grover PJ, Hutchinson PJ, Coulter IC. British Neurosurgical Trainee Research Collaborative. The management and outcome for patients with chronic subdural hematoma: A prospective, multicenter, observational cohort study in the United Kingdom. Journal of Neurosurgery. 2017;**127**(4):732–739. DOI: 10.3171/2016.8.JNS16134

[5] Tanaka T, Kaimori M. Histological study of vascular structure between the dura mater and the outer membrane in chronic subdural hematoma in an adult. No Shinkei Geka. Neurological Surgery. 1999;**27**(5):431–436

[6] Davies JM, Knopman J, Mokin M, Hassan AE, Harbaugh RE, Khalessi A, Fiehler J, Gross BA, Grandhi R, Tarpley J, Sivakumar W, Bain M, Crowley RW, Link TW, Fraser JF, Levitt MR, Chen PR, Hanel RA, Bernard JD, Jumaa M, Youssef P, Cress MC, Chaudry MI, Shakir HJ, Lesley WS, Billingsley J, Jones J, Koch MJ, Paul AR, Mack WJ, Osbun JW, Dlouhy K, Grossberg JA, Kellner CP, Sahlein DH, Santarelli J, Schirmer CM, Singer J, Liu JJ, Majjhoo AQ, Wolfe T, Patel NV, Roark C, Siddiqui AH. EMBOLISE Investigators Adjunctive Middle Meningeal Artery Embolization for Subdural Hematoma. New England Journal of Medicine. 2024;**391**(20):1890–1900. DOI: 10.1056/NEJMoa2313472

[7] Gillespie CS, Veremu M, Cook WH, Ashraf M, Lee KS, Chedid Y, Alam AM, Karepov Y, Davies BM, Edlmann E, Papanagiotou P, Korfias S, Santarius T, Minett T, Hutchinson PJ, Kolias A. Middle meningeal artery embolization for chronic subdural hematoma: Meta-analysis of three randomized controlled trials and review of ongoing trials. Acta Neurochirurgica. 2025;**167**(1):166. DOI: 10.1007/s00701-025-06587-4

[8] Fiorella D, Monteith SJ, Hanel R, Atchie B, Boo S, McTaggart RA, Zauner A, Tjoumakaris S, Barbier C, Benitez R, Spelle L, Pierot L, Hirsch JA, Froehler M, Arthur AS. STEM investigators. embolization of the middle meningeal artery for chronic subdural hematoma. New England

Journal of Medicine. 2025;**392**(9):855–864. DOI: 10.1056/NEJMoa2409845

[9] Link TW, Boddu S, Paine SM, Kamel H, Knopman J. Middle Meningeal Artery Embolization for Chronic Subdural Hematoma: A Series of 60 Cases. Neurosurgery. 2019;**85**(6):801–807. DOI: 10.1093/neuros/nyy521

[10] Ironside N, Nguyen C, Do Q, Ugiliweneza B, Chen CJ, Sieg EP, James RF, Ding D. Middle meningeal artery embolization for chronic subdural hematoma: A systematic review and meta-analysis. Journal of NeuroInterventional Surgery. 2021;**13** (10):951–957. DOI: 10.1136/neurintsurg-2021-017352

[11] Ganz JC. The lucid interval associated with epidural bleeding: Evolving understanding. Journal of Neurosurgery. 2013;**118**(4):739–745. DOI: 10.3171/2012.12.JNS121264

[12] Dossani RH, Kalakoti P, Thakur JD, Ommaya NAAK. Legacy and contributions to neurosurgery. Neurosurgery. 1930-2008;**80**(2):324–330. DOI: 10.1093/neuros/nyw031

[13] Mandai S, Sakurai M, Matsumoto Y. Middle meningeal artery embolization for refractory chronic subdural hematoma. Case Report. Journal of Neurosurgery. 2000;**93** (4):686–688. DOI: 10.3171/jns.2000.93.4.0686

[14] Ban SP, Hwang G, Byoun HS, Kim T, Lee SU, Bang JS, Han JH, Kim CY, Kwon OK, Oh CW. Middle meningeal artery embolization for chronic subdural hematoma. Radiology. 2018;**286**(3):992–999. DOI: 10.1148/radiol.2017170053

[15] Putnam TJ, Cushing H. Chronic subdural hematoma: Its Pathology, Its relation to pachymeningitis hemorrhagica and its surgical treatment. Archives of Surgery, Published Online. 1925;**11** (3):329–393. DOI: 10.1001/archsurg.1925.01120150002001

[16] Weigel R, Hohenstein A, Schilling L. Vascular endothelial growth factor concentration in chronic subdural hematoma fluid is related to computed tomography appearance and exudation rate. Journal of Neurotrauma. 2014;**31** (7):670–673. DOI: 10.1089/neu.2013.2884

[17] Stanisic M, Lyngstadaas SP, Pripp AH, Aasen AO, Lindegaard KF, Ivanovic J, Ilstad E, Konglund A, Sandell T, Ellingsen O, Saehle T. Chemokines as markers of local inflammation and angiogenesis in patients with chronic subdural hematoma: A prospective study. Acta Neurochirurgica. 2012;**154**(1):113–120. DOI: 10.1007/s00701-011-1203-2

[18] Ito H, Saito K, Yamamoto S, Hasegawa T. Tissue-type plasminogen activator in the chronic subdural hematoma. Surgical Neurology. 1988;**30** (3):175–179

[19] Nagahori T, Nishijima M, Takaku A. Histological study of the outer membrane of chronic subdural hematoma: Possible mechanism for expansion of hematoma cavity. No Shinkei Geka. Neurological Surgery. 1993;**21**(8):697–701

[20] Qiao Y, Alkarawi S, Provasek VE, Zhang YJ, Tsappidi S, Hui F. Chronic subdural hematoma: What precisely are we treating? Interventional Neuroradiology.

2024;15910199241263633. DOI: 10.1177/15910199241263633

[21] Siddiq F, Shakir M, Nguyen TN, Hassan AE, Abdalkader M, Kenmuir CL, Liebeskind DS, Novakovic R, Majidi S, Sheth SA, El-Ghanem M, Ortega-Gutierrez S, Guerrero WR, Malik AM. Consensus statement on middle meningeal artery embolization in chronic subdural hematoma treatment: A guideline from the society of vascular and interventional neurology guidelines and practice standards committee. Vascular and Interventional Neurology. 2025;**5**(6). DOI: 10.1161/SVIN.125.001814

[22] Scerrati A, Visani J, Ricciardi L, et al. To drill or not to drill, that is the question: Nonsurgical treatment of chronic subdural hematoma in the elderly. A systematic review. Neurosurgical Focus. 2020;**49**(4):E7. DOI: 10.3171/2020.7.FOCUS20237

[23] Matsumoto H, Hanayama H, Okada T, Sakurai Y, Minami H, Masuda A, Tominaga S, Miyaji K, Yamaura I, Yoshida Y. Which surgical procedure is effective for refractory chronic subdural hematoma? Analysis of our surgical procedures and literature review. Journal of Clinical Neuroscience. 2018;**49**:40–47. DOI: 10.1016/j.jocn.2017.11.009

[24] Azab MA, Sarhan K, Hernández-Hernández A, Atallah O, Clervius H. The many faces of middle meningeal artery: A comprehensive review of anomalous variations and endovascular implications. The Neuroradiology Journal. 2026;**39**(1):42–50. DOI: 10.1177/19714009251347828

[25] Shotar E, Meyblum L, Premat K, Lenck S, Degos V, Grand T, Cortese J, Pouvelle A, Pouliquen G, Mouyal S, Boch AL, Carpentier A, Sourour NA, Mathon B, Clarençon F. Middle meningeal artery embolization reduces the post-operative recurrence rate of at-risk chronic subdural hematoma. Journal of NeuroInterventional Surgery. 2020;**12**(12):1209–1213. DOI: 10.1136/neurintsurg-2020-016048

[26] Srivatsan A, Mohanty A, Nascimento FA, Hafeez MU, Srinivasan VM, Thomas A, Chen SR, Johnson JN, Kan P. Middle meningeal artery embolization for chronic subdural hematoma: Meta-analysis and systematic review. World Neurosurgery. 2019;**122**:613–619. DOI: 10.1016/j.wneu.2018.11.167

[27] Kim E. embolization therapy for refractory hemorrhage in patients with chronic subdural hematomas. World Neurosurgery. 2017;**101**:520–527. DOI: 10.1016/j.wneu.2017.02.070

[28] Ye S, Zhang S, Li T, Wang S, Wang X, Deng Y, Yu L, Li Q, Feng X, Jiang B, Dai Z, Guo Q, Han Q, Jin W, Weng S, Jiang S, Qiu Z, Abdalkader M, Nguyen TN. Middle meningeal artery embolization in the treatment of chronic subdural hematoma: A systematic review and meta-analysis. Cerebrovascular Diseases. 2026;**55**(1):125–135. DOI: 10.1159/000546001

[29] Chaliparambil RK, Jahromi BS, Metcalf-Doetsch W, Potts MB. Middle meningeal artery embolization for chronic subdural hematoma: A bibliometric analysis. World Neurosurgery. 2025;**193**:1145–1164. DOI: 10.1016/j.wneu.2024.10.136

[30] Chen H, Salem MM, Colasurdo M, Sioutas GS, Khalife J, Kuybu O, Carroll KT, Hoang AN, Baig AA, Salih M, Khorasanizadeh M, Baker C, Mendez

Ruiz A, Cortez GM, Abecassis Z, Ruiz Rodríguez JF, Davies JM, Narayanan S, Cawley CM, Riina H, Moore J, Spiotta AM, Khalessi A, Howard BM, Hanel RA, Tanweer O, Tonetti D, Siddiqui AH, Lang M, Levy EI, Jovin TG, Grandhi R, Srinivasan VM, Levitt MR, Ogilvy CS, Jankowitz B, Thomas AJ, Gross BA, Burkhardt JK, Kan P. Standalone middle meningeal artery embolization versus middle meningeal artery embolization with concurrent surgical evacuation for chronic subdural hematomas: A multicenter propensity score matched analysis of clinical and radiographic outcomes. Journal of NeuroInterventional Surgery. 2024;**16**(12):1313–1319. DOI: 10.1136/jnis-2023-020907

[31] Kan P, Maragkos GA, Srivatsan A, et al. Middle meningeal artery embolization for chronic subdural hematoma: A multicenter experience of 154 consecutive embolizations. Neurosurgery. 2021;**88**(2):268–277. DOI: 10.1093/neuros/nyaa432

[32] Salem MM, Kuybu O, Nguyen Hoang A, Baig AA, Khorasanizadeh M, Baker C, Hunsaker JC, Mendez AA, Cortez G, Davies JM, Narayanan S, Cawley CM, Riina HA, Moore JM, Spiotta AM, Khalessi AA, Howard BM, Hanel R, Tanweer O, Levy EI, Grandhi R, Lang MJ, Siddiqui AH, Kan P, Ogilvy CS, Gross BA, Thomas AJ, Jankowitz BT, Burkhardt JK. Middle meningeal artery embolization for chronic subdural hematoma: Predictors of Clinical and Radiographic Failure from 636 embolizations. Radiology. 2023;**307**(4):e222045. DOI: 10.1148/radiol.222045

[33] Mortezaei A, Al-Saidi N, Yahyaei K, Abdelsalam A, Guada Delgado L, Hanna J, Dmytriw AA, Rahmani R, Abla AA, Starke RM. Middle meningeal artery embolization versus standard of care for chronic subdural hematoma: Meta-analysis of randomized controlled trials. The Neuroradiology Journal. 2025;19714009251389579. DOI: 10.1177/19714009251389579

[34] Schmolling ÁH, Pérez-García C, Trejo C, López-Frías A, Jaroenngarmsamer T, Rosati S, Arrazola J, Moreu M. Middle meningeal artery embolization for management of chronic subdural hematoma. Radiographics. 2024;**44**(4):e230158. DOI: 10.1148/rg.230158

[35] Dzaye O, Brahmbhatt A, Abajian A, Moussa AM, Kkh Y, Moss NS, Newman WC, Lis E, Tabar V, Cornelis FH. Middle meningeal artery embolization using cone-beam computed tomography augmented guidance in patients with cancer. Diagnostic and Interventional Imaging. 2023;**104**(7-8):368–372. DOI: 10.1016/j.diii.2023.03.007

[36] Yang K, Kim KH, Lee HJ, Jeong EO, Kwon HJ, Kim SH. Role of adjunctive tranexamic acid in facilitating resolution of chronic subdural hematoma after surgery. Journal of Korean Neurosurgical Society. 2023;**66**(4):446–455. DOI: 10.3340/jkns.2022.0200

[37] Cohen-Cohen S, Jabal MS, Rinaldo L, Savastano LE, Lanzino G, Cloft H, Brinjikji W. Middle meningeal artery embolization for chronic subdural hematoma: A single-center experience and predictive modeling of outcomes. The Neuroradiology Journal. 2024;**37**(2):192–198. DOI: 10.1177/19714009231224431

[38] Hennessy MJ, Mathias JL, Guazzo EP. Improving the integration of health outcomes after chronic subdural haematoma: A rapid systematic review. Journal of Clinical Neuroscience. 2026;**143**:111713. DOI: 10.1016/j.jocn.2025.111713

Chapter 7

Transcallosal Versus Endoscopic Resection of Third Ventricular Colloid Cysts: A Contemporary Evidence-Based Comparison

Helen Fowler

Abstract

The management of colloid cysts of the third ventricle continues to evolve as neurosurgeons seek the optimal balance between surgical efficacy, patient safety, and minimally invasive intervention. This chapter critically appraises the two predominant operative modalities, the transcallosal microsurgical and endoscopic approaches, drawing upon contemporary literature, meta-analyses, and practice-based evaluations to establish evidence-informed recommendations. Historically, microsurgical transcallosal excision has been considered the gold standard due to its high rate of gross total resection (GTR); however, advancements in endoscopic instrumentation, neuronavigation, and intraoperative imaging have redefined the surgical landscape. Endoscopic resection now offers comparable success with reduced morbidity, shorter hospital stays, and improved patient recovery profiles. Despite this, variations in surgical preference, outcome reporting, and grading systems continue to hinder consensus. This chapter synthesises data from over a decade of comparative studies to delineate the benefits and limitations of each approach, with particular emphasis on operative technique, resection grading, and recurrence patterns. The introduction of structured frameworks such as the modified Colloid Cyst Risk Score (CCRS) is examined as a pivotal tool for guiding practice and improving interinstitutional consistency. In light of these developments, the chapter advocates for a standardised, evidence-based protocol integrating technological precision with individualised patient factors. By contextualising both the historical and emerging paradigms of colloid cyst surgery, this work contributes to the broader discourse on modern neurosurgical innovation and procedural refinement.

Keywords: colloid cyst, third ventricle, endoscopic neurosurgery, transcallosal approach, neuronavigation, image-guided surgery, gross total resection, intracranial pressure, cerebrospinal fluid, minimally invasive neurosurgery, surgical outcomes, recurrence, evidence-based practice

1. Introduction

Colloid cysts of the third ventricle are benign, epithelium-lined lesions arising from the roof of the third ventricle, near the Foramen of Munro. Although they account for only around 2% of intracranial tumours, their anatomical location results in obstruction of cerebrospinal fluid (CSF) pathways, potentially precipitating acute hydrocephalus or sudden neurological deterioration [1, 2]. Despite their non-aggressive histology, the risk of rapid clinical decline makes timely diagnosis and management essential [3, 4].

Histologically, colloid cysts consist of a thin epithelial lining overlying connective tissue, typically arising as developmental malformations that may remain asymptomatic for many years [2, 5]. Patients often present only when the cyst enlarges sufficiently to obstruct CSF flow, with headaches being the most common symptom, reported in over 60% of patients. Without intervention, progressive ventricular enlargement can lead to worsening symptoms and, in severe cases, sudden death [4, 6, 7].

There is currently no universally accepted standardised surgical protocol for colloid cyst management [8–10]. The most commonly performed approaches include the microsurgical transcallosal route, the transcortical route, and neuro-endoscopic excision [11]. Over the past three decades, the microsurgical transcallosal and neuroendoscopic techniques have emerged as the predominant approaches, each aiming to achieve durable resection whilst minimising neurological morbidity.

Endoscopic removal has advanced substantially since its introduction in the 1980s, supported by improvements in optics, stereotactic guidance, and minimally invasive instrumentation [12, 13]. Nevertheless, concerns regarding incomplete capsule removal and higher recurrence rates have historically led many surgeons to favour microsurgical techniques [14]. Both microsurgical and endoscopic techniques have evolved in parallel, driven by improvements in neuronavigation accuracy, refined retractor systems, and the availability of smaller microsurgical instruments that permit narrower operative corridors [14].

The evolution of surgical practice reflects an ongoing effort to balance the completeness of excision with minimisation of morbidity. The following section summarises comparative evidence on outcomes, recurrence, complications, and recent developments in grading and decision-making frameworks.

2. Anatomy and pathophysiology

The ventricular system produces and circulates CSF, which cushions the brain, maintains biochemical homoeostasis, and facilitates waste clearance [15]. CSF is generated primarily by the choroid plexus within the lateral, third, and fourth ventricles at an approximate rate of 500 ml per day [5]. From the lateral ventricles, CSF flows through the interventricular foramina of Monro into the third ventricle, before passing caudally through the cerebral aqueduct to the fourth ventricle. It then exits via the foramina of Luschka and Magendie into the subarachnoid space, where it circulates around the brain and spinal cord before being reabsorbed into the venous system [15, 16].

Colloid cysts arise in the anterior third ventricle, typically at or near the foramen of Monro, placing them in a critical position to obstruct CSF flow. As a cyst enlarges, even small increases in volume can impede CSF passage through the foramen, leading to obstructive hydrocephalus and raised intracranial pressure (ICP). This mechanism explains why some patients present with progressive symptoms, whereas others may deteriorate abruptly [5, 7]. Untreated obstruction can result in rapid neurological decline or sudden death, emphasising the need for timely surgical intervention when symptomatic or when radiological evidence of CSF compromise is present [5, 7].

Radiological assessment plays a central role in diagnosis and risk stratification. Traditional measures, such as the Evans Index, provide only a limited estimate of ventricular enlargement but correlate poorly with the degree of CSF obstruction or the likelihood of acute deterioration [17]. Contemporary imaging approaches, such as the assessment of ventricular asymmetry and foramen of Monro patency, offer more meaningful insights into the functional impact of the lesion and help guide the urgency of intervention [18, 19].

3. Historical evolution of surgical techniques

Operative approaches for third ventricular colloid cysts have historically varied, often reflecting the preferences, training, and experience of treating surgeons. This variability has contributed to inconsistencies in practice and, at times, unequal treatment pathways across patient groups [1, 20]. The evolution of surgical techniques must, therefore, be understood within the broader context of technological advancement, anatomical constraints, and the ongoing effort to balance surgical exposure with the preservation of neurological function.

The proximity of colloid cysts to eloquent cortical and subcortical structures has long necessitated meticulous surgical planning to minimise functional morbidity [21]. Early image-guided neurosurgical techniques, pioneered by Sir Victor Horsley and Robert Clarke, were limited by their inability to account for individual anatomical variability. These early systems assumed uniformity across patients, underscoring the requirement for patient-specific imaging to guide safe operative trajectories [22]. The introduction and widespread availability of CT and MRI in the 1980s represented a major advance, enabling clear differentiation between grey and white matter and providing the anatomical fidelity required for accurate navigation [22]. Building on these advances, neuronavigation has become integral to colloid cyst surgery, offering real-time intraoperative localisation and supporting minimally invasive approaches. Over the past three decades, this technology has evolved into an indispensable component of modern neurosurgical practice [14, 21, 22]

Within this technological landscape, the microsurgical transcallosal approach has emerged as a reliable and anatomically direct route to the third ventricle. Utilising a midline interhemispheric corridor, it allows access to the cyst while preserving critical structures such as the optic nerve and pituitary stalk [23]. Unlike neuroendoscopic procedures, which may be performed using frameless systems, microsurgical transcallosal approaches typically require rigid three-pin fixation to ensure stability and precision throughout the operation, preventing intraoperative head movement [10].

Operative approaches are varied and often dependent upon the treating surgeon at the time, giving rise to potential inequalities and treatment bias among patient groups [1]. Both microsurgical transcallosal and neuroendoscopic surgical approaches are established, yet debate persists regarding the optimal technique [9, 11]. Rapid deterioration can occur in untreated patients, underscoring the importance of timely intervention [11]. Complete excision of the cyst, including its capsule, is widely regarded as the most effective means of preventing recurrence [9, 11, 24].

Sampath et al. [25] emphasise that each surgical approach carries distinct advantages and limitations. The ideal procedure would combine short operative duration, high rates of gross total resection (GTR), low complication rates, and brief hospitalisation; however, no single technique consistently fulfils all criteria. As a result, surgical decision-making must balance patient-specific anatomy, cyst morphology, surgeon expertise, and institutional capability. This nuanced interplay between anatomical constraints, technological evolution, and clinical judgement has shaped the historical trajectory of colloid cyst surgery and continues to inform contemporary practice.

4. Comparative analysis

Contemporary literature comparing transcallosal and endoscopic resection demonstrates a complex interplay between surgical exposure, extent of resection, perioperative morbidity, recurrence patterns, and the influence of evolving operative technologies. Although the evidence base is shaped largely by retrospective series, multi-centre reviews, administrative datasets, and systematic analyses, several consistent themes emerge. Microsurgical transcallosal resection continues to demonstrate the highest rates of GTR and the lowest recurrence, reflecting the advantages of direct visualisation and unrestricted operative access [1, 4, 7, 23].

In contrast, neuroendoscopic surgery offers a distinctly different benefit: reduced morbidity, shorter hospitalisation, and faster post-operative recovery, particularly in patients with favourable ventricular anatomy [9, 26, 27]. These divergent strengths highlight the fundamental challenge in colloid cyst surgery: balancing the durability of resection with the imperative to minimise operative risk and providing the context for a more detailed exploration of how anatomical factors, technological capability, and surgeon expertise influence outcomes across both operative strategies. See **Table 1**, which summarises GTR rates, recurrence, morbidity, and hospital stay from all major cited comparative studies.

4.1 Extent of resection and recurrence

Multiple studies consistently demonstrate that microsurgical transcallosal excision provides the most reliable means of capsule removal, with meta-analyses reporting GTR rates approaching 97% [7]. This advantage reflects the wide operative exposure and direct visualisation afforded by microsurgery, allowing meticulous dissection of adherent capsule from surrounding neurovascular structures. In contrast, although neuroendoscopic series have shown progressive improvement over time, complete capsule excision remains more challenging within the constraints of a narrow operative corridor. Contemporary endoscopic GTR rates of approximately 58% reflect these anatomical limitations, even in experienced centres [14, 28].

Author	GTR rates	Recurrence and reoperation	Morbidity and mortality	Hospital stay
Boogaarts et al. [17]	57.5% near complete removal; 42.5% residual cyst	6 required repeated endoscopic surgery for regrowth.	Permanent morbidity noted in 1 patient	Not reported
Sribnick et al. [24]	100% reported complete excision	1 recurrence reported	Mortality reported 3 deaths (x 1 during the perioperative period, x 1 at 46 days post-surgery, and x 1 not related). Persistent memory loss reported in 6 patients.	Median length: 5 days (range: 2–63).
Sheikh, Mendelson and Liu [7]	Microsurgical: 96.8% GTR Endoscopic: 58.2% GTR	Recurrence – microsurgical 1.48% vs. endoscopic 3.91% reoperation – microsurgical 0.38% vs. endoscopic 3%	Morbidity Microsurgical 16.3% Endoscopic 10.5% Mortality showed no significant difference: microsurgical 1.4% vs. endoscopic 0.6%.	Not reported
Brostigen et al. [29]	GTR was observed in 82% of microsurgical patients and 62% in endoscopic approaches; however, 7 endoscopic approaches were converted to microsurgical due to difficulties.	6.3% underwent repeat surgery (one due to residual, one due to recurrence).	All post-operative complications occurred in the patients undergoing endoscopic procedures (13%, n = 4). No surgical mortality was noted.	Not reported
Connolly et al. [9]	Not discussed.	Readmission was reported in 9.6% of endoscopic patients and 17.3% of microsurgical patients. Reoperation was observed in 0.85% (n = 2) of microsurgical patients and 0.41% (n = 1) of endoscopic patients.	Complications were noted in 48.6% of microsurgical patients vs. 44.3% of endoscopic patients. 2 mortalities were reported in microsurgical patients (0.83%), and one mortality was reported in endoscopic patients (0.41%).	Longer hospital stays are reported with microsurgical approaches, in addition to higher costs.
Sethi et al. [11]	Microsurgical GTR 96.1% Endoscopic GTR 78.5%	Microscopic recurrence 0.98% Endoscopic recurrence 4.57% Reoperation Microsurgical 0.33% Endoscopic 0.61%	Mortality reported in 0.65% for microsurgical and 1.06% for endoscopic. Overall morbidity was significantly higher in the microsurgery group – 18.6% vs. 8.7%.	Not reported
Isaacs et al. [18]	Did not discuss GTR 98.6% of patients experienced complete emptying of cyst contents with capsule coagulation.	91.9% had no recurrence. Recurrence was reported in 6 patients; 5 underwent secondary surgery.	No mortality or morbidity reported, other than two minor complications: a post-operative seizure (in a known epileptic) and a wound infection.	Not reported

Author	GTR rates	Recurrence and reoperation	Morbidity and mortality	Hospital stay
Heller and Heilman [14]	Cyst wall completely removed in 14/15 microsurgical cases and 3/13 endoscopic cases. Cyst wall remnants were left in 11 cases.	Recurrence was reported in 8 patients, with 6 of them having undergone the original endoscopic surgery. Surgery was required in 7 of these 8 cases. Recurrence was noted in 55% of patients where a cyst wall remnant was left behind.	1 patient from microsurgical approach developed transient leg weakness, 1 a bone infection and 1 a UTI. 1 patient, having undergone endoscopic resection, was readmitted twice. No mortality was discussed.	Length of hospital stay averages 3.8 days for microsurgical cases and 3.5 days for endoscopic cases.
Alkhaibary et al. [1]	N/A	Recurrence Microsurgical 9.1% Endoscopic 14.3%	Complications were reported in 4 microsurgical patients and 3 endoscopic patients.	Microsurgical 20.6 (SD 14.7) Endoscopic 14.3 days (SD 8.7)
Sayehmiri et al. [26]	Microsurgical 98.15% Endoscopic 91.29%	Minimal recurrence was reported in microsurgical cases, compared to 1.78% reported in endoscopic cases. Reoperation Microsurgical 0.09% Endoscopic 0.49%	Complications were reported in 20.68% of microsurgical patients, compared to 10.42% of endoscopic patients. Mortality reported as 0.18% in microsurgical cases and 0.00% in endoscopic cases.	Microsurgical reports average 7.85 days Endoscopic reports average 4.69 days.
Farahmand et al. [27]	Not discussed	Not discussed	Not discussed	Microsurgical cases: 11.3 days (SD: 10.4). Endoscopic cases: 5.6 days (SD: 6.1).
Ajlan et al. [10]	GTR reported in 66% (n = 6) Near total resection reported 33% (n = 4)	No patients required reoperation or experienced any reported recurrence.	Reported 1 patient developing supplementary motor area syndrome, which resolved at 1 year follow-up. 1 reported memory disturbance 1 reported hemiplegia 1 reported post-op seizures All of the above were resolved between discharge and follow-up.	Not discussed.
Unal et al. [4]	GTR reported in 81% (n = 17) Subtotal resection reported in 14% (n = 3) Partial resection reported in 5% (n = 1)	No recurrences were reported in the follow-up period (2–24 months, mean follow-up period: 14 months).	1 reported case of permanent hemiplegia. 1 reported case of meningitis. No mortality reported	Not reported

Table 1.
Summary of GTR rates, recurrence, morbidity, and hospital stay from all major cited comparative studies.

Recurrence is strongly associated with residual capsule, irrespective of the surgical modality used. This principle has been repeatedly demonstrated across microsurgical and endoscopic series and remains one of the most robust predictors of long-term outcomes. The Colloid Cyst Extent of Resection Grade (CcERG) grading system, proposed by Isaacs et al. [18], quantifies the extent of capsule removal, correlating it with recurrence risk. This highlights that recurrence clusters predominantly in cases with less than 50% capsule removal (Grade III), whereas patients with more extensive resection, whether achieved endoscopically or microsurgically, experience markedly lower recurrence rates. This finding reinforces the importance of cyst wall clearance, rather than operative modality alone, and highlights the importance of documenting the rationale for any intentional remnant [18].

Further studies support this graded relationship, demonstrating that, when complete evacuation of cyst contents is achieved, partial capsule removal beyond a critical threshold may still offer durable disease control [4, 28]. The threshold for 'adequate' resection remains up for debate, and the heterogeneity of reporting across series complicates attempts to define an international, universal standard. What is clear is that the extent of capsule removal must be interpreted within the context of ventricular anatomy, cyst orientation, and intraoperative safety. In some cases, particularly those with dense vascular adherence, attempting full excision may introduce disproportionate risk, making a graded, anatomy-dependent approach more appropriate.

4.2 Morbidity and functional outcomes

The principal limitation of microsurgery is its higher morbidity. Studies consistently report increased rates of memory impairment, venous infarction, seizures, meningitis, and prolonged hospitalisation following microsurgical transcallosal resection [11, 26, 27, 29]. Neuroendoscopic surgery, by contrast, is associated with shorter operative times, reduced neurological complications, and faster recovery. Large administrative datasets further demonstrate higher post-operative seizure rates and 30-day readmissions within microsurgical cohorts, contributing to increased healthcare costs [9]. Endoscopic cohorts generally experienced more favourable immediate post-operative recovery and higher patient-reported satisfaction [24]. Connolly et al. [9] identified significantly higher post-operative seizure rates and 30-day readmissions in microsurgical patients, with complications imposing substantially greater healthcare costs.

4.3 Hydrocephalus and anatomical considerations

Anatomical factors play a decisive role in determining the safest and most effective surgical approach for third ventricular colloid cysts. Enlarged ventricles generally favour endoscopic access, whereas complex cyst morphology, vascular adherence, or limited visualisation may necessitate microsurgical conversion [9, 16, 23, 24]. Risk-mapping studies further demonstrate that cysts located within specific ventricular 'zones' carry differing risks of hydrocephalus and acute neurological deterioration, providing valuable guidance for triage and urgency of treatment [20]. Importantly, hydrocephalus itself does not reliably predict the rate of clinical decline, although it is more frequently observed in symptomatic patients. Studies such as

Farahmand et al. [27] reiterate that obstructive CSF physiology, rather than cyst size alone, drives acute morbidity, underscoring the need for timely, complete removal in selected patients.

4.4 Technological evolution

Technological advancement has played a central role in shaping contemporary outcomes for both microsurgical and endoscopic resection of third ventricular colloid cysts. Although the most dramatic improvements have occurred within neuroendoscopy, driven by enhanced optics, angled instruments, improved navigation systems, and advances in surgeon experience, microsurgical transcallosal techniques have also evolved. Refinements in operative microscopes, retractor systems, and neuronavigation integration have contributed to the consistently high rates of GTR reported in modern microsurgical series [14, 21, 30, 31].

Across multiple decades of literature, these technological developments are reflected in progressively improved endoscopic outcomes. Early endoscopic series, conducted during the developmental phase of the technique, often reported lower resection rates due to limited visualisation and less sophisticated equipment. In contrast, contemporary studies by Boogaarts et al., Isaacs et al., and Sethi et al. all document markedly improved performance, with GTR rates approaching 80% in experienced centres [11, 17, 18]. These refinements, combined with greater operator familiarity, help explain the variability observed between older and more recent series. They also highlight the importance of interpreting pooled historical data with caution, as differences in equipment, technique, and training can significantly influence reported outcomes.

4.5 Methodological limitations

Despite the thematic consistency across the literature, the methodological limitations of existing studies significantly constrain the strength of comparative conclusions. The literature is limited by its retrospective design, heterogeneous definitions of GTR, and single-centre series with long inclusion periods, during which operative techniques, equipment, and surgeon experience evolved substantially. These inconsistencies introduce unavoidable selection and performance biases that complicate direct comparisons between microsurgical and endoscopic cohorts.

A further challenge lies in the heterogeneity of outcome reporting. Definitions of GTR vary widely, particularly within endoscopic studies, where some authors equate complete resection with full capsule removal, while others accept coagulation or partial excision [11, 18]. This inconsistency undermines the reliability of pooled analyses and contributes to the apparent variability in recurrence rates across series. Follow-up duration is similarly inconsistent, with many studies lacking long-term imaging surveillance capable of detecting late recurrence. Attrition, incomplete datasets, and variable reporting of complications further limit the interpretability of published outcomes.

Technological evolution adds an additional layer of complexity. As highlighted in the preceding section, improvements in endoscopic optics, angled instruments, and navigation systems have markedly enhanced resection capability over time. Microsurgical techniques have also benefited from incremental refinements in operative microscopes, retractor systems, and the integration of neuronavigation. These advancements mean that older series may not reflect contemporary practice, and pooled historical data risk obscuring the true performance of modern techniques.

Taken together, these methodological constraints underscore the need for caution when interpreting comparative outcomes and highlight the importance of prospective, standardised research. Without uniform definitions of resection, consistent imaging protocols, and reporting frameworks, meaningful comparison between surgical approaches remains challenging. Addressing these limitations will be essential for developing more robust evidence to guide future clinical decision-making.

5. Synthesis and evidence-based recommendations

Current evidence supports an individualised, anatomy-dependent strategy when selecting between microsurgical transcallosal and neuroendoscopic resection of third ventricular colloid cysts. No single operative modality is universally superior; instead, outcomes depend on the interplay between ventricular size, cyst morphology, vascular relationships, and the technical constraints of each approach. Endoscopic surgery is most appropriate for patients with enlarged ventricles, favourable cyst orientation, minimal vascular adherence, and obstructive hydrocephalus, where the advantages of a minimally invasive, low-morbidity intervention can be fully realised. In contrast, microsurgical transcallosal resection remains preferable when complete capsule excision is essential, particularly in young patients, those with large or multi-lobulated cysts, or cases in which endoscopic visualisation is restricted. Microsurgery also retains a central role in the management of recurrence following an initial endoscopic procedure and in patients with small ventricles or unfavourable anatomy, where the narrow operative corridor limits endoscopic manoeuvrability.

A key determinant of long-term cure is the extent of capsule removal. Microsurgical exposure typically facilitates full excision, which correlates with the lowest recurrence rates across the literature. However, several contemporary endoscopic series suggest that removal of at least 50% of the capsule may sufficiently reduce recurrence risk, provided that the cyst contents are fully evacuated and the remnant is carefully coagulated. This emerging evidence supports a more nuanced interpretation of 'adequate' resection in endoscopic practice, although it also underscores the importance of meticulous intraoperative documentation. Surgeons should clearly record the rationale for any intentional remnant, as post-operative surveillance and patient counselling must be tailored to the degree of residual tissue.

Given the potential for intraoperative challenges, including vascular adherence, bleeding, or impaired visibility, conversion from endoscopy to microsurgery should remain an available and readily deployable option. This requirement reinforces the principle that centres offering endoscopic treatment must maintain immediate microsurgical capability and ensure that operative planning incorporates contingency pathways [14]. Such preparedness is essential to avoid compromising resection quality or patient safety when endoscopic limitations are encountered.

Post-operative imaging surveillance should be proportionate to the extent of resection achieved. Patients with complete or near-complete capsule removal may be followed with MRI at six months, one year, and subsequently every two to three years, depending on the treating clinician [32]. In contrast, those with any radiological remnant require more frequent monitoring, given the documented potential for late recurrence. This stratified approach aligns surveillance intensity with recurrence risk and ensures timely detection of regrowth.

Finally, the literature consistently highlights the need for standardised reporting practices, particularly regarding the definition of gross total resection. Variability in how GTR is defined and documented has hindered meaningful comparison across studies and contributed to ongoing controversy regarding the optimal surgical approach. Adoption of structured resection-grading tools, such as the CcERG, offers a practical means of reducing interinstitutional variability and improving the reliability of outcome reporting [20]. Establishing such standardisation is essential for advancing evidence-based practice, refining patient selection, and enabling more robust multi-centre collaboration. **Figure 1** demonstrates a decision-making flowchart based on the individualised strategy discussed in this section.

6. Operative time and hospital stay

Operative duration consistently differs between approaches, with endoscopic procedures typically requiring significantly less time than microsurgical resections [33]. Shorter operative duration is associated with reduced hospital stays, smaller incisions, and potentially lower infection rates. Cost analyses demonstrate that post-operative complications, which are more common following longer microsurgical procedures, substantially increase healthcare expenditure [20].

Length of hospital stay similarly favours endoscopic surgery, with multiple studies reporting shorter post-operative admissions compared with microsurgical groups. Given increasing healthcare pressures, even modest reductions in hospitalisation carry meaningful benefits [9]. These findings support the use of endoscopy as a first-line option in suitable patients, reserving microsurgery for anatomically complex or recurrent cases [1, 26].

7. Recurrence and complications

Cyst size and anatomical configuration exert a significant influence on both recurrence risk and the choice of surgical approach. While some authors advocate endoscopy for smaller lesions, others argue that cysts without ventriculomegaly or those positioned unfavorably, particularly when access corridors are narrow, may be more safely addressed via microsurgical routes. These divergent recommendations highlight a broader issue within the field: reliance on individual surgeon preference rather than standardised, evidence-based criteria. Such variability inevitably contributes to inconsistent patient outcomes and complicates attempts to compare results across institutions.

A major challenge in interpreting recurrence data is the lack of a consistent definition of GTR. Endoscopic studies vary widely, with some defining complete resection as the removal of the entire cyst, while others consider coagulation of the capsule sufficient. This inconsistency contributes to heterogeneous outcome

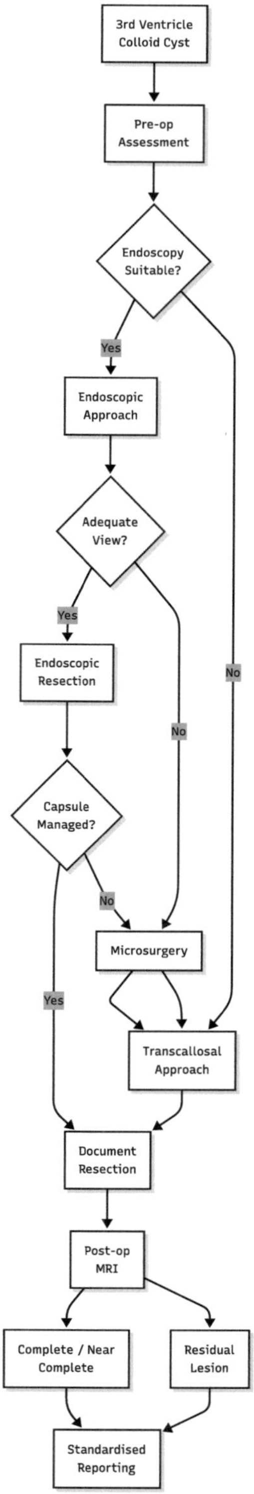

Figure 1.
Decision-making flow chart.

reporting data and reflects the absence of a universally recognised gold standard management strategy [11, 18].

However, reliance on intraoperative surgical judgement alone has limitations. As demonstrated in evaluations of Simpson grading for meningiomas, modern post-operative imaging frequently provides a more reliable assessment of resection completeness than the surgeon's impression [19]. Applying this principle to colloid cyst surgery suggests that a combined model, integrating the CcERG grading system with routine post-operative MRI, may offer a more accurate and contemporary method for evaluating surgical adequacy. Importantly, such a system could be applied uniformly across both endoscopic and microsurgical approaches, addressing the current lack of standardisation within the literature.

Overall, while microsurgical excision consistently achieves higher rates of complete capsule removal and correspondingly lower recurrence, it carries greater morbidity due to the need for brain retraction and more extensive operative exposure [7]. Endoscopic techniques provide a less invasive alternative with reduced post-operative morbidity, but recurrence risk increases when GTR is not achieved [1]. These trade-offs underscore the importance of aligning the surgical approach with individual anatomical and clinical factors rather than relying solely on surgeon preference.

Standardising definitions of resection, incorporating structured grading tools, and ensuring long-term imaging follow-up will be essential for refining recurrence estimates and guiding future surgical decision-making. Such measures would not only improve comparability across studies but also support the development of more consistent, evidence-based treatment pathways for patients with third ventricular colloid cysts.

8. Conclusion and future directions

Early evaluations of endoscopic outcomes were conducted during the developmental phase of neuroendoscopy, when operative techniques, optical quality, and documentation standards were still evolving. Contemporary clinical practice reflects this progress: modern endoscopic platforms, improved visualisation, and routine integration of neuronavigation have enhanced precision and reproducibility. Consequently, recent studies show increasing comparability between endoscopic and microsurgical outcomes, particularly regarding the extent of resection.

However, the wider evidence base remains limited by small sample sizes, short and inconsistent follow-up, variable definitions of GTR, and selection bias. These factors continue to constrain meaningful cross-study comparisons and impede definitive treatment recommendations. Given the rarity of colloid cysts, most available evidence derives from retrospective, single-centre series spanning long inclusion periods. Although valuable, these studies have restricted generalisability and should be interpreted within the context of institutional expertise, demographic variation, and resource availability. Nonetheless, the existing literature provides coherent themes and exposes clear gaps that offer a meaningful foundation for more rigorous future investigation.

Progress will depend on coordinated, prospective, multi-centre research programmes capable of systematically comparing endoscopic and microsurgical approaches using standardised definitions of resection and uniform post-operative imaging protocols. Achieving adequately powered studies will require broad interinstitutional collaboration and extended recruitment periods, but the potential benefits, including validated grading systems, clearer treatment pathways, and improved long-term outcomes, justify these efforts.

The challenges associated with researching rare neurosurgical conditions are evident throughout this chapter. Limited patient numbers, varied methodological approaches, and inconsistent reporting standards make it difficult to draw firm comparisons or establish consensus across studies. Rather than viewing these constraints as barriers, they highlight the importance of embedding research activity within routine clinical practice. Strengthening institutional audit processes, fostering collaboration between centres, and ensuring structured long-term follow-up will be essential for building a more reliable evidence base. By integrating these practices into everyday care, the field can gradually overcome the limitations imposed by rarity and continue to refine the management of third ventricular colloid cysts.

Looking ahead, the future of colloid cyst surgery lies in the integration of precision, personalisation and standardisation. Continued refinement of minimally invasive techniques, wider adoption of structured resection-grading systems, and consistent long-term imaging follow-up will be essential to improving outcome reliability across centres. Equally important is the development of collaborative research networks and shared outcome registries capable of generating prospective data. By aligning operative strategy with patient-specific anatomy and embedding standardised reporting frameworks into routine practice, the field can move toward clearer treatment pathways, greater reproducibility, and more personalised operative decision-making. Through these collective efforts, neurosurgical teams will be better positioned to deliver safe, precise, and durable care for patients with third ventricular colloid cysts.

9. Search strategy

Several medical research databases were utilised with the criteria below, including EMBASE, CINAHL, Ovid, Scopus, Medline, and PubMed.

Person – adults

Intervention – endoscopic approach for the excision of a colloid cyst

Comparison – transcallosal microsurgical approach for the excision of a colloid cyst

Observation – post-operative resection; morbidity and mortality; length of hospital stay; complication rates; reoperation rates.

Inclusion criteria – adult; colloid cyst; English language; endoscopic; transcallosal

Exclusion criteria – paediatric; cyst aspiration alone; transventricular; non-English language

Search terms – 'transcallosal', 'endoscopic', 'colloid cyst', 'transcortical', as key words alone and in combination, in addition to article titles, were screened for relevant abstracts, and those that met the criteria were then included. Backward chaining was also used to ensure the inclusion of studies that had not appeared in the original search.

Selection criteria – articles were included if they reported gross total resection, recurrence rates, mortality, hospital stay, complication rates, and shunt dependency. Papers from 2010 onwards were favoured as they are more representative of the operative techniques currently in use compared to previous, dated techniques, in addition to permitting a higher quality comparison of outcomes through the use of contemporaneous studies.

Those papers deemed pertinent and reliable, following a review of abstracts and key words used, as well as the elimination of any duplicates, have been included within this review.

Acknowledgements

I would like to express my heartfelt appreciation to Professor Michael Jenkinson, an eminent expert in the field of neurosurgery. It has been an honour to work alongside you and to witness your exceptional skill, precision, and steadfast commitment to improving patient outcomes. Thank you for your wit, your many moments of laughter – often at my accent – and your constant understanding and encouragement throughout our shared clinical experiences.

My sincere thanks also go to Mr. Jawad Yousaf, whose enthusiasm, humour, and pioneering spirit in advancing endoscopic surgery have been truly inspiring. Your dedication to innovation and your unwavering focus on delivering the highest standards of patient care have profoundly influenced my own professional development.

It has been an honour to learn from and work alongside such remarkable individuals, both professionally and personally. Your influence has profoundly shaped not only my clinical growth but also my resilience and dedication to the field of neurosurgery and academia.

Conflict of Interest

The author declares no conflicts of interest.

Disclosure

The author acknowledges the use of Microsoft Copilot solely as a supportive tool for refining the organisational structure, clarity, and flow of this chapter. Copilot was not used to generate, interpret, or introduce scientific content, nor was it involved in the development of any clinical analysis, conceptual frameworks, or literature synthesis presented within the body of this work. All substantive material originates entirely from the author's own expertise, professional experience, and independent review of the scientific sources cited.

The author confirms that no AI-generated factual data, references, or clinical claims have been incorporated into this work and that all literature used has been sourced manually through recognised academic channels. Responsibility for the accuracy, originality, and integrity of the chapter remains exclusively with the author, in accordance with IntechOpen's policies on authorship, accountability, and permissible use of AI-assisted technologies.

Author details

Helen Fowler
Liverpool John Moores University, Liverpool, England

*Address all correspondence to: H.L.Fowler@ljmu.ac.uk

References

[1] Alkhaibary A, Baydhi L, Alharbi A, Alshaikh AA, Khairy S, Abbas M, et al. Endoscopic versus open microsurgical excision of colloid cysts: A comparative analysis and state-of-the-art review of neurosurgical techniques. World Neurosurgery. 2021;**149**:e298–e308

[2] Roberts A, Jackson A, Bangar S, Moussa M. Colloid cyst of the third ventricle. Journal of the American College of Emergency Physicians Open. 2021;**2**(4):e12503. DOI: 10.1002/emp2.12503

[3] Indiran V, Mohamed MFS, Santhanam R, Maduraimuthu P. Colloid cyst causing varying obstructive hydrocephalus. Turkish Journal Of Neurology. 2018;**2018**(1):79–80. DOI: 10.4274/tnd.46514

[4] Unal TC, Sencer A, Dolas I, Gulsever CI, Sahin D, Dolen D, et al. Full-endoscopic removal of third ventricular colloid cysts: Technique, results, and limitations. Frontiers in Surgery. 2023;**10**. DOI: 10.3389/fsurg.2023.1174144

[5] Lindsay KW, Bone I, Fuller G, 5th Ed. Neurology and neurosurgery illustrated. Edinburgh: Churchill Livingstone/Elsevier; 2016

[6] Hernesniemi J, Romani R, Dashti R, Albayrak BS, Savolainen S, Ramsey C, et al. Microsurgical treatment of third ventricular colloid cysts by interhemispheric far lateral transcallosal approach—experience of 134 patients. Surgical Neurology. 2008;**69**(5):447–453. DOI: 10.1016/j.surneu.2007.11.005

[7] Sheikh AB, Mendelson ZS, Liu JK. Endoscopic versus microsurgical resection of colloid cysts: A systematic review and meta-analysis of 1278 patients. World Neurosurgery. 2014;**82**(6):1187–1197. DOI: 10.1016/j.wneu.2014.06.024

[8] Melicher D, Gaál S, Berényi T, Fenyves BG, Novák PK, Tóth A, et al. Acute hydrocephalus caused by a colloid cyst — A case report. International Journal of Emergency Medicine. 2023;**16**(1):28. DOI: 10.1186/s12245-023-00500-5

[9] Connolly ID, Johnson E, Lamsam L, Veeravagu A, Ratliff J, Li G. Microsurgical vs. endoscopic excision of colloid cysts: An analysis of complications and costs using a longitudinal administrative database. Frontiers in Neurology. 2017;**9**:8. DOI: 10.3389/fneur.2017.00259

[10] Ajlan A, Alwadee R, Basindwah S, Alsabbagh B, Alhumud MH, Alaskar AM. Outcomes and complications of microscopic transcallosal colloid cyst resection: A single institution experience. Surgical Neurology International. 2023;**14**:88. DOI: 10.25259/SNI_1071_2022

[11] Sethi A, Cavalcante D, Ormond DR. Endoscopic versus microscopic transcallosal excision of colloid cysts: A systematic review in the era of complete endoscopic excision. World Neurosurgery. 2019;**132**:e53–8. DOI: 10.1016/j.wneu.2019.08.259

[12] Hellwig D, Bauer BL, Schulte M, Gatscher S, Riegel T, Bertalanffy H. Neuroendoscopic treatment for colloid cysts of the third ventricle. Neurosurgery. 2008;**62**(6):SHC1101–9.

DOI: 10.1227/01.NEU.0000333776.89187.0E

[13] Vorbau C, Baldauf J, Oertel J, Gaab MR, Schroeder HWS. Long-term results after endoscopic resection of colloid cysts. World Neurosurgery. 2019;**122**:e176–85. DOI: 10.1016/j.wneu.2018.09.190

[14] Heller RS, Heilman CB. Colloid Cysts: Evolution of surgical approach preference and management of recurrent cysts. Operative Neurosurgery. 2020;**18**(1):19–25. DOI: 10.1093/ons/opz059

[15] Stratchko L, Filatova I, Agarwal A, Kanekar S. The ventricular system of the brain: Anatomy and normal variations. Seminars in Ultrasound, CT and MRI. 2016;**37**(2):72–83. DOI: 10.1053/j.sult.2016.01.004

[16] Spennato P, Cinalli MA, Di Costanzo M, Vitulli F, Ruggiero C, Di Martino G, et al. Endoscopic ultrasonic aspirator–assisted removal of a third ventricular colloid cyst. Child's Nervous System. 2023;**39**(12):3391–3395. DOI: 10.1007/s00381-023-05994-1

[17] Boogaarts HD, Decq P, Grotenhuis JA, Le Guérinel C, Nseir R, Jarraya B, et al. Long-term results of the neuroendoscopic management of colloid cysts of the third ventricle: A series of 90 cases. Neurosurgery. 2011;**68**(1):179–187. DOI: 10.1227/NEU.0b013e3181ffae71

[18] Isaacs AM, Bezchlibnyk YB, Dronyk J, Urbaneja G, Yong H, Hamilton MG. Long-Term outcomes of endoscopic third ventricle colloid cyst resection: Case series with a proposed grading system. Operative Neurosurgery. 2020;**19**(2):134–142. DOI: 10.1093/ons/opz409

[19] Chotai S, Schwartz TH. The Simpson Grading: Is It Still Valid? Cancers (Basel). 2022;**14**(8):2007. DOI: 10.3390/cancers14082007

[20] Beaumont TL, Limbrick DD, Rich KM, Wippold FJ, Dacey RG. Natural history of colloid cysts of the third ventricle. Journal of Neurosurgery. 2016;**125**(6):1420–1430. DOI: 10.3171/2015.11.JNS151396

[21] Chartrain AG, Kellner CP, Fargen KM, Spiotta AM, Chesler DA, Fiorella D, et al. A review and comparison of three neuronavigation systems for minimally invasive intracerebral hemorrhage evacuation. Journal of NeuroInterventional Surgery. 2018;**10**(1):66–74. DOI: 10.1136/neurintsurg-2017-013091

[22] Galloway RL. Introduction and historical perspectives on image-guided surgery. In: Golby AJ, editor. Image-guided neurosurgery. Amsterdam: Elsevier; 2015. p. 1–22. DOI: 10.1016/B978-0-12-800870-6.00001-7

[23] Graziano F, Ganau M, Meccio F, Iacopino D, Ulm A. The transcallosal anterior interfoniceal approach: A microsurgical anatomy study. Journal of Neurological Surgery Part B: Skull Base. 2014;**76**(03):183–188. DOI: 10.1055/s-0034-1396595

[24] Sribnick EA, Dadashev VY, Miller BA, Hawkins S, Hadjipanayis CG. Neuroendoscopic colloid cyst resection: A case cohort with follow-up and patient satisfaction. World Neurosurgery. 2014;**81**(3–4):584–593. DOI: 10.1016/j.wneu.2013.12.006

[25] Sampath R, Vannemreddy P, Nanda A. Microsurgical excision of colloid cyst with favorable cognitive outcomes and short operative time and

hospital stay. Neurosurgery. 2010;**66**(2):368–375. DOI: 10.1227/01.NEU.0000363858.17782.82

[26] Sayehmiri F, Starke RM, Eichberg DG, Ghanikolahloo M, Rahmatian A, Fathi M, et al. Comparison of microscopic and endoscopic resection of third-ventricular colloid cysts: A systematic review and meta-analysis. Clinical Neurology and Neurosurgery. 2022;**215**:107179. DOI: 10.1016/j.clineuro.2022.107179

[27] Farahmand D, Stridh J, Ziegelitz D, Tisell M. Endoscopic versus open microsurgery for colloid cysts of the third ventricle. British Journal of Neurosurgery. 2023;**37**(1):59–62. DOI: 10.1080/02688697.2021.1925872

[28] Gurses ME, Lu VM, Gecici NN, Shah KH, Gökalp E, Bashti M, et al. Utilizing tubular retractors in colloid cyst resection: A single surgeon experience. Surgical Neurology International. 2024;**15**:179

[29] Brostigen CS, Meling TR, Marthinsen PB, Scheie D, Aarhus M, Helseth E. Surgical management of colloid cyst of the third ventricle. Acta Neurologica Scandinavica. 2017;**135**(4):484–487. DOI: 10.1111/ane.12632

[30] Khoshnevisan A, Allahabadi NS. Neuronavigation: Principles, clinical applications and potential pitfalls. Iranian Journal of Psychiatry. 2012;7(2):97–103

[31] Golby AJ. Image-guided Neurosurgery. Amsterdam: Academic Press (Elsevier); 2015

[32] D'Souza WP, Sarkar S, Chacko G, Rajshekhar V. Recurrence after microsurgical excision of third ventricular colloid cysts in a contemporary cohort of 84 patients. World Neurosurgery. 2025;**194**:123421. DOI: 10.1016/j.wneu.2024.11.004

[33] Alkhaibary A, Baydhi L, Alharbi A, Alshaikh AA, Khairy S, Abbas M. Endoscopic versus open microsurgical excision of colloid cysts: A comparative analysis and state-of-the-art review of neurosurgical techniques. World Neurosurgery. 2021;**149**:e298–308. DOI: 10.1016/j.wneu.2021.02.032

Chapter 8

Deep Brain Stimulation and MRI-Guided Focused Ultrasound for Movement Disorders – Neurosurgical Considerations and Institutional Experiences

Bea Drucks, Hubertus Maximilian Mehdorn and Isabel Luebbing

Abstract

Neurosurgical therapy of movement disorders experienced a renaissance with the introduction of deep brain stimulation (DBS) in the 1980s. DBS aimed to improve symptoms in Parkinson's disease, tremor, and dystonia, which are among the most important indications. With advances in brain MRI and the implementation of real-time intracerebral temperature monitoring, focused ultrasound has again expanded treatment options by broadening the neurosurgical therapeutic armamentarium. Having started DBS in 1999 and having contributed extensively to the scientific literature together with our neurological colleagues, we have, since 2020, also used MR-guided focused ultrasound (MRgFUS) to treat patients with tremor by targeting the ventral intermediate nucleus (VIM; nucleus ventralis intermedius) and patients with Parkinson's disease by targeting the subthalamic nucleus (STN; nucleus subthalamicus). Differential indications for DBS and MRgFUS will be reviewed, and experiences from a high-volume center with more than 1,500 patients treated by either method will be described.

Keywords: movement disorders, Parkinson's disease, tremor, deep brain stimulation, MR-guided focused ultrasound, differential indications

1. Introduction

Neurodegenerative diseases, such as Parkinson's disease, have increasingly been described over the last centuries, particularly more recently in aging populations in Western countries [1]. With increasing knowledge of pathophysiology [2] and grading of clinical features [3], therapies that evolved gradually – both medical and surgical – could be evaluated scientifically. In this manuscript, the evolution of surgical therapy is described based on personal experience, with a focus on deep brain stimulation (DBS) and MR-guided focused ultrasound (MRgFUS).

1.1 Short history of functional neurosurgery

Neurosurgical therapy for severe pain conditions and movement disorders such as Parkinson's disease, tremor, dystonia, as well as psychiatric disorders, has developed into stereotactic functional neurosurgery as part of an expanding neurosurgical subspecialty. Initially, lesioning of deep-seated brain areas during open surgical procedures was common, often in the form of lobotomy. Zernov and Altukhov in Moscow used an "encephalometer" in epilepsy patients in 1891, which they had presented a few years earlier; Horsley and Clarke developed a stereotactic apparatus that was later refined for clinical application by Spiegel and Wycis [4, 5].

In 1947, they presented a stereotactic system firmly attached to the skull to reliably target deep-seated brain areas and to create lesions using electrodes or cryoprobes, while controlling the effects of lesioning in the awake patient [6].

The technique was based on knowledge of the functions of deeply located brain areas, on clinical observation of symptom improvement (e.g., after stroke in patients with Parkinson's disease), and on open surgical destruction of fiber tracts, as well as on the reproducible relationship of relevant structures to the ventricular system – particularly the third ventricle – as an anatomical coordinate system. In their monograph, Spiegel and Wycis [7] describe technical details of this stereotactic method, including specialized probes/electrodes, and also discuss the problem of incorrectly placed lesions. Correct placement required careful calculation using the X, Y, and Z axes [4, 8, 9].

Highly consistent anatomical correlations between the midcommissural plane (defined as the midline between the anterior and posterior commissures within the third ventricle) and relevant basal ganglia nuclei, such as the globus pallidus internus (GPi), the subthalamic nucleus (STN), and the ventral intermediate nucleus (VIM), were recognized. Translating anatomical data into a clinically and surgically useful system required a stereotactic coordinate system and its representation in stereotactic atlases. Initially, these were paper atlases, for example, Spiegel and Wycis, Schaltenbrand [10], and others [4, 8, 9] and later computer-based atlases [11].

In clinical practice, a surgical coordinate system is traditionally based on a frame attached to the patient's skull, for example, most recently using a carbon ring (Inomed, Germany), as used in our department to allow MRI scanning (see **Figure 2**). Over time, frameless stereotactic systems became available. While initially the accuracy of frameless stereotactic systems was questioned, later meta-analyses showed that the difference was clinically insignificant [12].

In the 1960s and 1970s, the refinement of pharmacological therapy and the introduction of L-dopa to improve symptoms of Parkinson's disease led to a sharp decline in lesioning surgery, which subsequently focused more on pain treatment than on the treatment of movement disorders. While electrophysiology during these treatments was initially used to better target the area to be lesioned, A. Benabid in Grenoble [13] and J. Siegfried in Zurich (whom the senior author had the pleasure of observing during surgery in 1973) realized in the 1980s that continuous high-frequency stimulation (above 100 Hz), initially of areas of the thalamus and later the subthalamus, could improve symptoms of tremor and symptoms of Parkinson's disease. By systematically exploring the options of continuous stimulation, improving the understanding of the underlying pathophysiology as excitatory and inhibitory circuits [14] in a delicately balanced system, and advancing both hardware and electrophysiological management, Benabid and his group pioneered the particular field of DBS for movement disorders. Later, the indication for DBS was expanded to include psychiatric and other disorders [15, 16].

While the GPi was initially used for Parkinsonian symptoms and the VIM for tremor-dominant patients, the smaller STN subsequently became a preferred target for DBS in patients with Parkinson's disease in many centers [17].

1.2 Technological advances

Parallel to the improvement of neurosurgical technology per se, major advances in imaging technology – nowadays considered indispensable – gradually became available for depicting intracranial anatomy: from air-filled ventriculography using X-ray, to high-quality cranial computed tomography (cCT), and to magnetic resonance imaging (MRI), which enabled high-quality 3D insights into brain anatomy relevant for DBS, including functional imaging [18, 19].

The latest development in MR imaging leads to real-time intracerebral temperature measurement during the application of focused ultrasound to perform defined thermal lesions in stereotactically defined areas. Improvements in understanding (patho)physiology of basal ganglia lead to further improvement of treatment results.

Modern technologies have dramatically improved our ability to perform, with reasonably good success rates, such delicate treatments, which will be described in detail in the following pages. Nevertheless, neurosurgical therapy of brain disorders – whether more or less invasive – should be applied to our patients using the highest ethical and technical standards. In our opinion, this requires a dedicated team approach within a high-volume center that provides various treatment options.

While modern technologies have improved our ability to treat patients with movement disorders, psychiatric problems, pain, and other conditions using DBS, it should be remembered that fundamental knowledge of functional brain anatomy is derived from lesioning surgery. Much of the previously accumulated knowledge and experiences of lesioning seem to have diminished in high-income countries due to a variety of reasons, which should not be discussed further. In order to promote the idea of transferring old knowledge and implementing as much modern technology (e.g., imaging and electrophysiological mapping) as reasonable, a number of teaching courses in a variety of countries have been organized [20] to keep the options of lesioning open if stimulation techniques are not possible or not available. In the Asian hemisphere, an international society has even been founded.

At the same time, emerging neurosurgical indications and treatment options should be evaluated in multicenter, interdisciplinary studies with long-term perspectives in order to further optimize them and enable the best possible treatment for our patients, including socioeconomic points of view.

Some of the studies in which our department has participated are listed in the references [21–25]. Similarly, focused ultrasound for movement disorders is also being evaluated scientifically in a number of international studies in which our center participates.

2. Kiel neurosurgical experiences

2.1 DBS – general considerations

DBS is a surgical procedure that has been routinely performed at our institution since 1999, with a current operative volume of between 50–60 patients per year (see **Figure 1**). The major indications include Parkinson's disease, essential tremor, and dystonia; rarer indications at our center include epilepsy, Meige syndrome, Huntington's disease, and SUNCT (shortlasting unilateral neuralgiform headache with conjunctival injection, tearing, sweating and rhinorrhea), among others.

During DBS, electrodes are placed with millimeter precision into defined target regions of the brain using stereotactic techniques. Details of our techniques were published early on and have evolved further [26]. Planning of the target point and the electrode trajectory is based on high-resolution, high-precision MRI. The electrodes are connected via thin extension cables, which are tunneled subcutaneously to an implantable pulse generator (IPG) positioned below the clavicle. Stimulation parameters of the pulse generator can and must subsequently be adjusted transcutaneously according to the clinical symptoms.

The therapeutic effect of DBS is based on the modulation of neuronal networks [15, 27]. This means that the electrical impulses delivered by the neurostimulator influence the activity of the targeted brain regions, leading to an improvement in clinical symptoms. For example, hyperactive areas within specific nuclei can be inhibited without destruction of neuronal tissue [22]. Contrary to early assumptions that DBS merely suppresses certain brain regions, it is now understood to involve a complex interplay at both the cellular and neurochemical levels. Important mechanisms include the suppression of beta activity, which directly correlates with Parkinsonian symptoms [28]. In addition, DBS exerts effects at the chemical level through the release of specific neurotransmitters, which, in turn, modulate neuronal responses and thereby positively influence clinical symptomatology [29, 30].

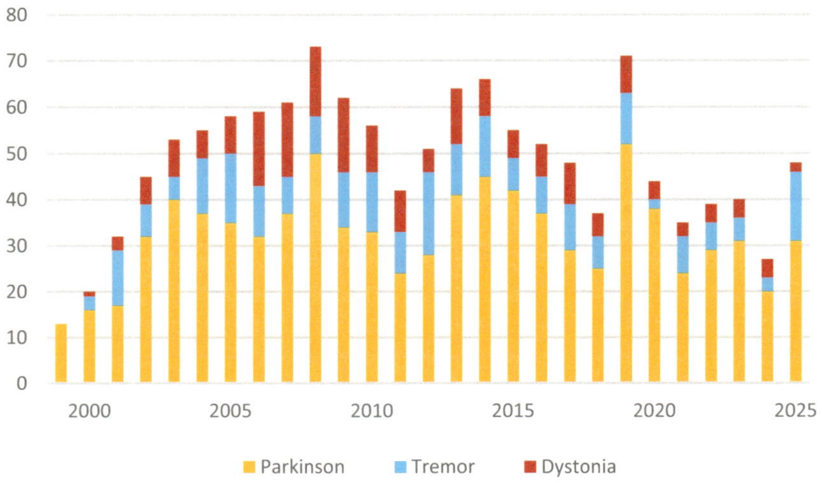

Figure 1.
Major DBS indications 1999–2025 total N = 1,301.

The indications for DBS are broad and, in addition to movement disorders, encompass mostly experimental conditions such as epilepsy, eating disorders, obsessive–compulsive disorder, Tourette syndrome, substance use disorders, depression, and chronic pain [13, 16, 25, 26, 31–37].

Careful patient selection remains a key determinant of long-term success following DBS. In Parkinson's disease, relevant considerations include a sustained response to levodopa, the burden of axial symptoms, cognitive status, and psychiatric comorbidity. In dystonia, outcomes are influenced by age at surgery, disease duration, fixed musculoskeletal deformities, and genetic subtypes (e.g., DYT1-positive dystonia), while secondary dystonia often shows more variable responses.

With respect to movement disorders, which constitute the primary focus of this chapter, several disease entities have been shown to respond favorably to DBS. These include Parkinson's disease with motor fluctuations, medication-induced dyskinesia, and pharmacotherapy-refractory tremor. Regulatory approvals in the USA include essential tremor and Parkinson's disease (1997), dystonia (2003), obsessive–compulsive disorder (2009), and epilepsy (2018). Atypical and secondary parkinsonian syndromes are excluded from the patient cohort, as no clear therapeutic benefit has been demonstrated in these populations.

Essential tremor represents another established indication, typically presenting in a severe, functionally disabling form that is insufficiently controlled by pharmacological treatment. Furthermore, primary generalized dystonia, as well as cervical and segmental dystonia – particularly in cases of inadequate response to medical therapy – can be treated effectively with DBS.

In general, contraindications to DBS include comorbid conditions that preclude surgical eligibility, as well as coagulation disorders that cannot be adequately corrected. In patients with Parkinson's disease, psychiatric comorbidities and cognitive decline, including dementia, constitute additional contraindications. From a neurosurgical perspective, contraindications are primarily related to target and trajectory planning and may include advanced cerebral atrophy, anatomical variants of vascular anatomy, vascular malformations, space-occupying lesions, and prior intracranial interventions, any of which may preclude safe implantation.

Potential complications associated with DBS include intracerebral hemorrhage, which is rare and has been reported with a probability of approximately 1.6% per electrode [38]. These hemorrhages are typically small and asymptomatic. In addition, infections of the implanted hardware may occur, even at long intervals after surgery, necessitating partial or complete explantation of the system in some cases. In patients who have received long-term stimulation, system explantation with abrupt cessation of stimulation may be associated with a potentially life-threatening withdrawal syndrome. Other device-related complications include hardware fracture and/or lead dislocation. Possible stimulation-related side effects include dysarthria and gait instability, which can often be influenced by adjustment of stimulation parameters. Furthermore, preexisting psychiatric comorbidities may be aggravated.

At our center in Kiel, all patients with movement disorders are first seen by neurology for optimization of medical therapy. They are then evaluated in a specialized functional neurosurgical outpatient clinic to discuss surgical aspects of therapy. Each patient undergoes a comprehensive medical history assessment and a detailed neurological examination. Target selection and trajectory planning are performed using Brainlab neuronavigation software. Based on the aforementioned criteria, the indication for DBS is determined through an interdisciplinary consensus in collaboration with the Department of Neurology.

2.2 DBS – stepwise procedure

Patients are treated in a three-step procedure. Following a specialized MRI examination – performed on the day of surgery itself (1999–2015) or on the day prior to surgery – entry points, targets, and the intracranial trajectory are planned on a stereotactic workstation (Brainlab, Germany; Origin Server 3.3).

2.2.1 Step 1: Frame application

Under general anesthesia, the stereotactic ring/frame is applied. To verify exact intraoperative positioning and geometry, the use of the phantom method is advisable (see **Figure 2**).

2.2.2 Step 2: Lad implantation (awake, electrophysiology-guided)

An intraoperative CT scan is obtained and fused with the MRI datasets for stereotactic image guidance. Electrode implantation is then performed in the awake patient under local anesthesia. A segmented lead electrode is implanted into each designated brain region under continuous electrophysiological monitoring (see **Figure 3**). Recording electrophysiological activity while advancing the microelectrode helps define anatomical borders of the target area. Test stimulation through the microelectrode, while slowly withdrawing the microdrive, allows evaluation of the clinical effect of microstimulation. If the expected clinical benefit is not achieved or stimulation-induced side effects are pronounced, an alternative trajectory parallel to the central trajectory can be explored.

Stereotactic coordinates are defined as follows: the X coordinate corresponds to the mediolateral axis, the Y coordinate corresponds to the anteroposterior axis, and the Z coordinate corresponds to the superoinferior axis.

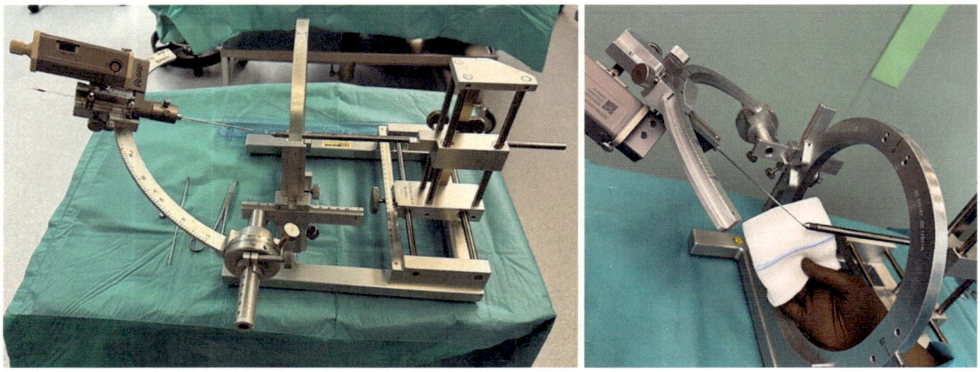

Figure 2.
Stereotactic system (Inomed, Germany): phantom with x, y, z coordinates simulating the target point. On the left, guide sleeve with motor drive (Alpha Omega, Germany) and microelectrode (Inomed). Careful control: the tip of the microelectrode is accurately positioned at the target point defined by the phantom simulation.

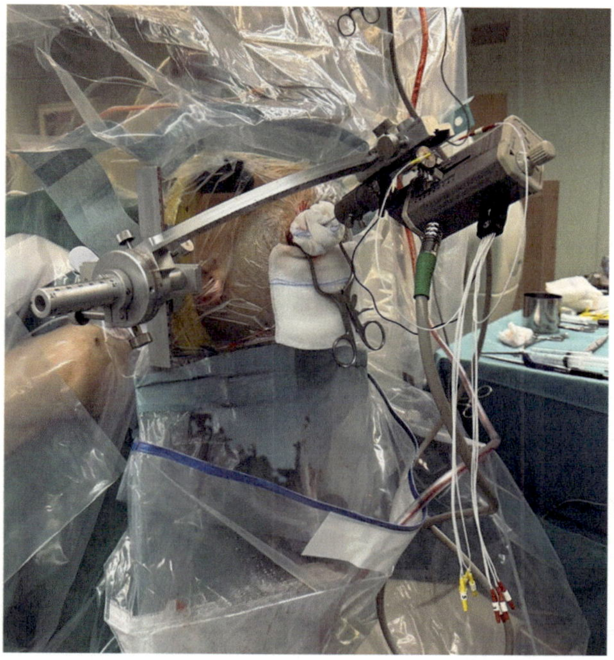

Figure 3.
The stereotactic system (Inomed) is mounted on the stereotactic frame (Inomed) with the guide sleeve, the motor drive (Alpha Omega) and the microelectrode. For electrophysiological recordings, the appropriate cables are attached to the recording electrode (red) and the test electrode (yellow).

After definitive lead placement, the patient undergoes a second CT scan to confirm the lead's location.

2.2.3 Step 3: IPG implantation

Implantation of the IPG is performed under general anesthesia. The IPG is typically positioned approximately two fingerbreadths inferior to the clavicle in a subcutaneous pocket, and the system is connected via subcutaneous extension cables.

Following radiological confirmation of correct electrode positioning (i.e., the postoperative CT control described above), the patient is observed in an intermediate care unit or regular ward. Stimulation can be initiated a few days later, once perilesional edema has decreased and the electrode placement (microlesion) effect has diminished.

2.3 MRgFUS

MRgFUS is an incisionless, noncraniotomy treatment modality that has been performed at our institution since 2020. Depending on regional regulatory frameworks and established clinical practice, the conditions considered approved/established for treatment include essential tremor, parkinsonian tremor, dystonia-associated tremor, Parkinson's disease, and neuropathic pain.

2.3.1 Treatment of movement disorders using MRgFUS

This treatment is considered an incisionless, noncraniotomy procedure, as it does not require opening of the skull. Yet one has to be aware of the disruptive/destructive nature of the procedure, although the intensity of the lesioning and its extent can be controlled in a highly reliable fashion under continuous imaging control.

MRgFUS exploits the ability to measure real-time conditions with the dedicated sequence of the 3T-MRI machine, enabling localized heating of up to 60°C at a predefined intracranial target. This warming is achieved by converging hundreds of ultrasound beams from the FUS system Exablate Prime (Insightec, Israel). This results in precise and permanent focused deactivation of neurons without causing permanent damage to the surrounding tissue, and the improvement of clinical symptoms can be analyzed on-site [39].

Depending on the treated condition, total treatment duration (including preparation and postprocedural care) typically ranges from approximately 1.5–3.5 hours. Shorter treatment times are common in tremor patients, whereas longer times may be required for tailored treatment of portions of the STN in patients with Parkinson's disease. The patient remains awake throughout the entire intervention to allow for repeated neurological testing.

Prior to treatment, complete shaving of the scalp is performed, followed by wet shaving immediately before the procedure. If a longer procedure is anticipated, a urinary catheter is placed preinterventionally. Anesthesiological standby is highly recommended in order to treat possible hypertensive events.

As a first step, local anesthesia is administered at the insertion points of the stereotactic frame, followed by its application (see **Figure 4a**). In order to allow homogeneous sonication, a membrane is placed watertight onto the skull so that the space between the sonication helmet and the skull surface can be filled with distilled water (see **Figure 4b**).

Next, the patient is positioned within the transducer of the focused ultrasound system, which is placed in the MRI scanner (see **Figure 5.a**). Particular care needs to be taken to ensure patient comfort throughout the procedure (**Figure 5.b**).

Following this step and system calibration, the actual treatment is performed with regular interim assessments of clinical efficacy. During the short sonication periods (typically between 11 and 30 seconds), increasing thermal exposure may cause transient headache, dizziness, or nausea, which are managed in collaboration with the attending anesthesiology team. In general, transient side effects such as paresthesias, gait instability, coordination disturbances, or speech impairment may occur [40]. With correct lesion positioning, these effects are usually transient, as they are due to perilesional edema, which will subside over the next few days. In order to reduce these complications, patients usually receive dexamethasone.

The risk of complications, such as intracerebral hemorrhage or permanent severe neurological deficits, has been reported to be low (e.g., <1% in reported clinical experience/series). Patients are typically discharged home or transferred to subsequent rehabilitation on the day following the intervention.

In principle, the movement-disorder indications are similar for both treatment modalities. Analogous to DBS, Parkinson's disease can be treated using three different targets: classical thalamotomy for tremor, subthalamotomy targeting the STN in Parkinson's disease, and pallidotomy [39, 41–43].

DOI: http://dx.doi.org/10.5772/intechopen.1015010

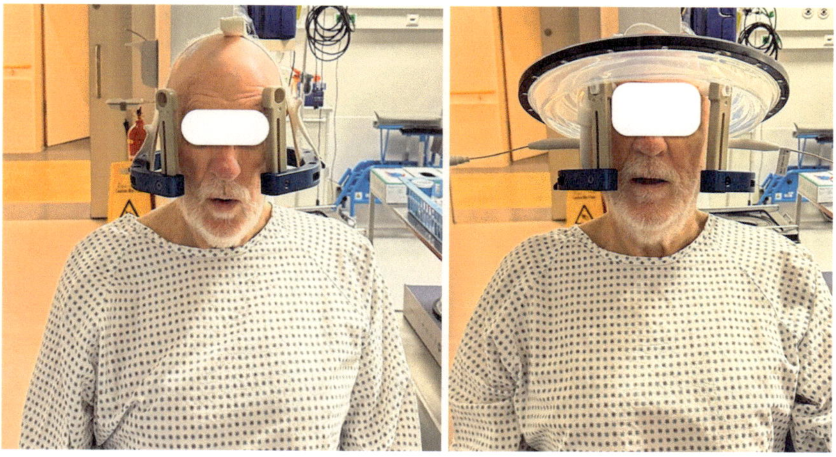

Figure 4.
(a) Positioning the reference frame; (b) adding the watertight membrane to the skull.

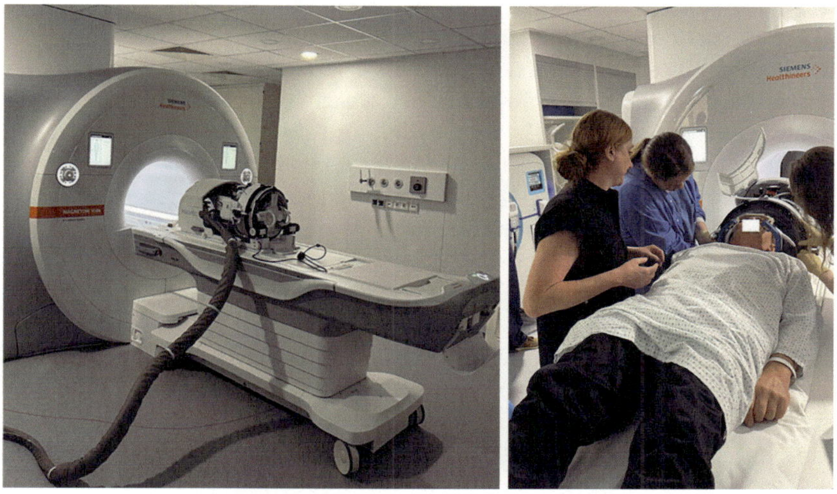

Figure 5.
(a) 3-Tesla MRI (Siemens Healthcare) with the connected Exablate Prime system. (b) Patient positioned with frame and membrane in the MRgFUS system (Exablate Prime, Insightec).

In the cases of thalamotomy for bilateral tremor, with initially very favorable responses on the more seriously affected side, the indication for contralateral treatment has recently emerged following an appropriate recovery period (a minimum of 6 months) and is currently under study.

Beyond movement disorders, some centers have started to treat neuropathic pain, analogous in principle to DBS.

Exclusion criteria include an unfavorable skull density ratio (SDR), for which different centers apply varying cutoff values [44]; in Kiel, based on experience accumulated over recent years, a range of 0.4–0.6 is used (see **Figure 6**).

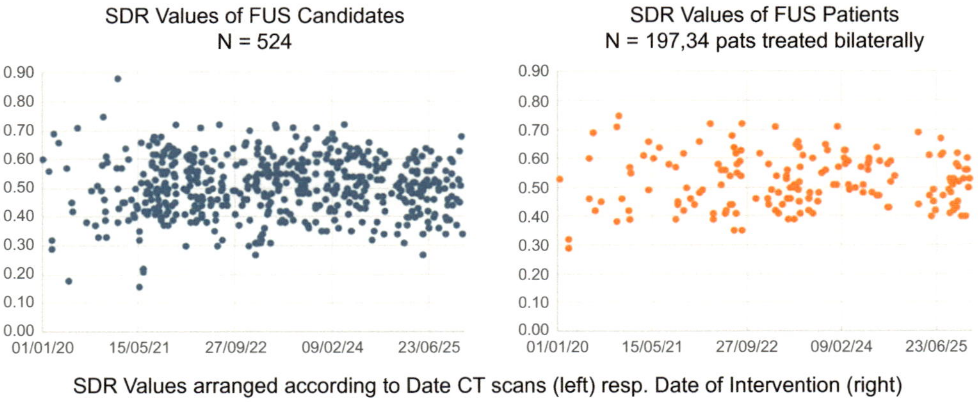

Figure 6.
Skull density ratio excludes possible candidates for FUS.

Contraindications: only patients who are eligible for MRI and who can tolerate the procedure without sedation in a static position for several hours can be treated. Any patient movement results in the inaccuracy of the target point, thereby negating the validity of meticulous millimeter-accurate planning.

Further, general relative contraindications – depending on the target location and similar to DBS – include intracranial tumors, vascular abnormalities, coagulation disorders, medication-resistant hypertension, and the use of drugs associated with an increased risk of bleeding that cannot be discontinued. Patients with psychiatric disorders or substance abuse may also be excluded from treatment with MRI-guided focused ultrasound, as cooperation during the procedure is mandatory.

2.3.2 MRgFUS stepwise procedure

Targeting follows the basic rules of stereotactic definition of targets mentioned above. The current standard is inspired by our cooperation with the Madrid MRgFUS center in Móstoles (Spain); most of their workflow is gratefully implemented in Kiel, and additional Kiel-specific procedural elements have been added. A strict protocol is followed, with the possibility of adjustment to sonication parameters.

Sonication intensity (typical temperature ranges):

- First step (alignement): 44–48°C
- Second step (verification): 49–52°C
- Third step (lesioning): 55–60°C
- Final step: repeated lesioning-level sonications until adequate symptom control is reached and the definitive lesion is created.

Sonication duration at approximately 60°C should generally not exceed 20 seconds; alternatively, two sonications of shorter duration within the 55–60°C range can be applied.

Step-by-step MRgFUS procedure

1. Preparation: after applying local anesthesia for positioning the Insightec stereotactic frame using sizable adapters, the patient lies on an MRI table, with the head secured in the stereotactic frame holder and the space between the sonication machine and the membrane filled with distilled water for ultrasound transmission.

2. Targeting and planning: real-time 3R MR images are used to identify the precise target location. Previous stereotactic planning in Brainlab is possible but not obligatory.

3. Low-energy test sonication: low-energy ultrasound is delivered to the test target location, checking for efficacy and side effects without causing irreversible damage.

4. High-energy treatment (ablation): once confirmed, high-intensity ultrasound energy is focused to heat the target tissue (57–63°C), creating a permanent lesion with perifocal edema.

5. Real-time monitoring: MRI continuously monitors temperature (thermography) and the location of the thermal focus, ensuring safety for surrounding tissues.

6. Final assessment and discharge: on the first day after treatment, posttreatment MR images (see **Figure 7**) verify that the target has been reached and that the intended lesion has been performed without side-effects; the patient is usually discharged on the same day.

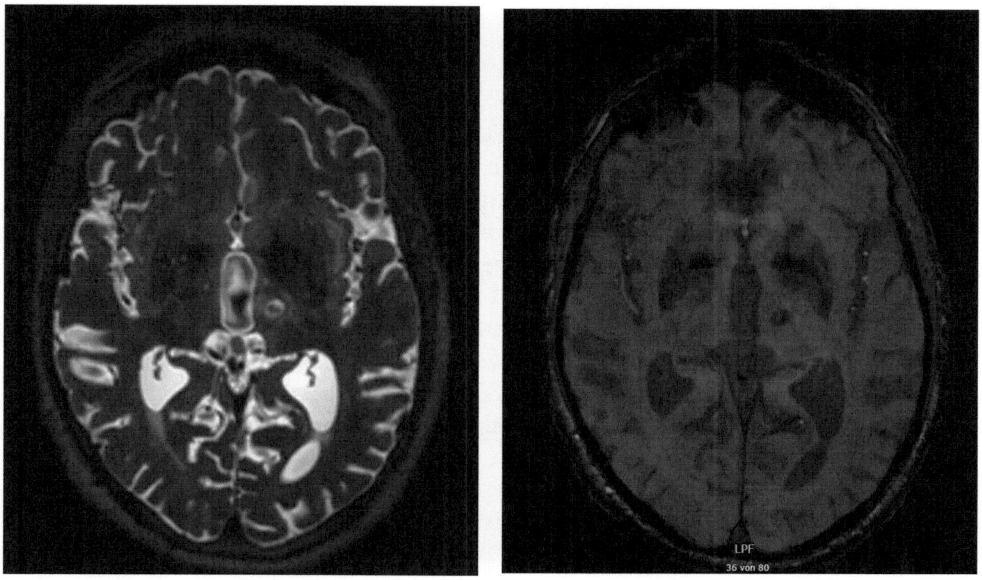

Figure 7.
Axial MRI performed 24 hours postthalamotomy (VIM) for ET. Left: t1-weighted sequence; right: SWI sequence. The lesion with adjacent edema in the left VIM nucleus is evident in both images.

Procedural details may undergo changes in the future, and further targets may be added.

3. Comparison of DBS and MRgFUS

Provided that the anatomical and clinical–neurological prerequisites are met (see above), individual patient preferences play a major role in treatment selection. Some patients decline surgical intervention due to the associated risks or because of the need for permanent electrical stimulation and implantation of foreign material. Other patients experience severe anxiety related to MRI, rendering MRI-guided focused ultrasound unsuitable, or, in the presence of bilateral symptoms, prefer the single-stage nature of bilateral DBS over staged MRgFUS.

A major advantage of MRgFUS is its ability to treat at least certain components of movement disorders without the need for invasive neurosurgical intervention and the associated procedural risks and complications. A technical limitation, however, is the lack of compensation for potential patient movement within the system, which may necessitate restarting the entire procedure, for example, in older patients with neck problems or when the need arises to void the bladder. In contrast, in DBS, the stereotactic frame in relation to the patient's head represents the primary reference; this relationship remains unchanged even in the event of patient movement, thereby reducing the risk of inaccuracy with respect to the planned target.

Another limitation of MRgFUS is that many disorders treated are progressive in nature, whereas ultrasound therapy provides symptom relief at a specific time point. Over subsequent years, symptoms may deteriorate, potentially necessitating repeated treatment. By comparison, DBS allows for targeted adaptation to symptom progression due to its therapeutic window, the availability of multiple stimulation programs, and the ability to stimulate different contacts, thereby modifying the electrical field.

Surgical treatment with DBS requires the continuous presence of anesthesiology personnel, as parts of the procedure are performed under general anesthesia with endotracheal intubation and require constant monitoring and adjustment of vital parameters. Some centers perform MRgFUS without anesthesiological support. Based on our experience in Kiel, however, many patients exhibit elevated blood pressure values prior to and during the intervention, in part due to the temporary discontinuation of maintenance medication, making pharmacological blood pressure management via continuous infusion necessary in many cases. During sonications, systolic blood pressure should not exceed 160 mmHg and should preferably be maintained at lower levels, as blood-brain barrier disruption may increase the risk of intralesional bleeding.

Overall, each modality has distinct advantages and limitations. Contrary to widespread claims, high-intensity focused ultrasound is not suitable for all patients with movement disorders. DBS remains a well-established, safe, and effective treatment modality with decades of clinical experience for the management of movement disorders, among other indications. Nevertheless, MRgFUS offers substantial and potentially groundbreaking opportunities for the treatment of a wide range of neurological conditions, particularly with regard to future expansions of indications, such as the transient opening of the blood–brain barrier.

When comparing both modalities, it becomes evident that they represent highly effective treatment options, particularly for patients with movement disorders, and, in many cases, allow for near-complete symptom control.

Figures 8–10 show the distribution of patients treated over the last five years at our center, since we have been able to offer both treatments to our patients.

As already recognizable in **Figure 1**, variations in treatment numbers may occur and may depend on many factors beyond the reach of neurosurgeons.

Whether the older age distribution observed in MRgFUS reflects the general health status and referral characteristics of patients presented to our center and considered for intervention warrants further study.

4. Socioeconomic aspects of DBS versus MRgFUS treatment

The costs of treatment are inherently dependent on the healthcare system of the country in which the intervention is performed. The data presented here, therefore, refer to the German healthcare system, which is comparable in structure to those of many other Western countries.

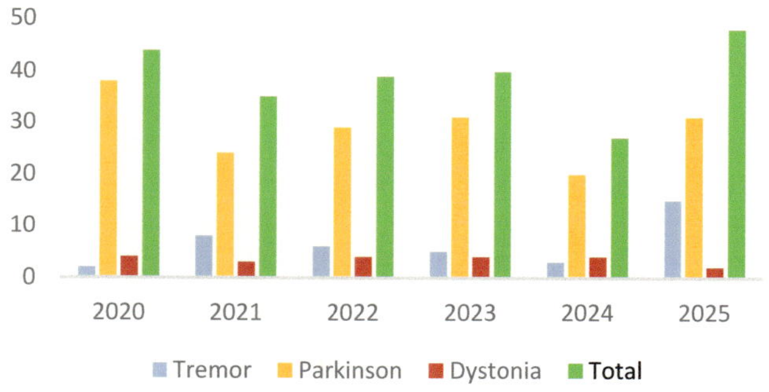

Figure 8.
Major DBS indications treated between 2020 and 2025 (N = 233).

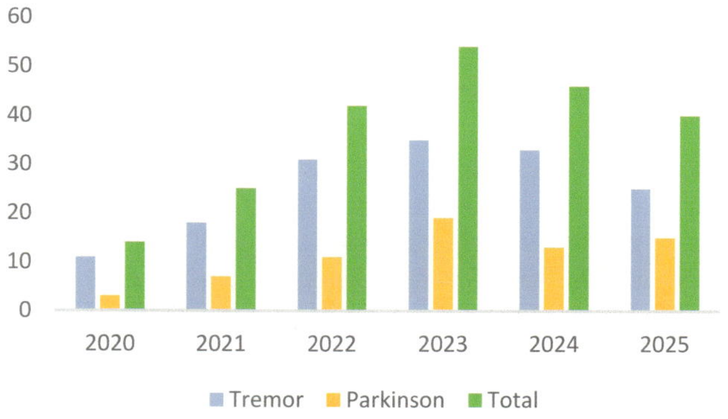

Figure 9.
MRgFUS indications treated between 2020 and 2025 (N = 221).

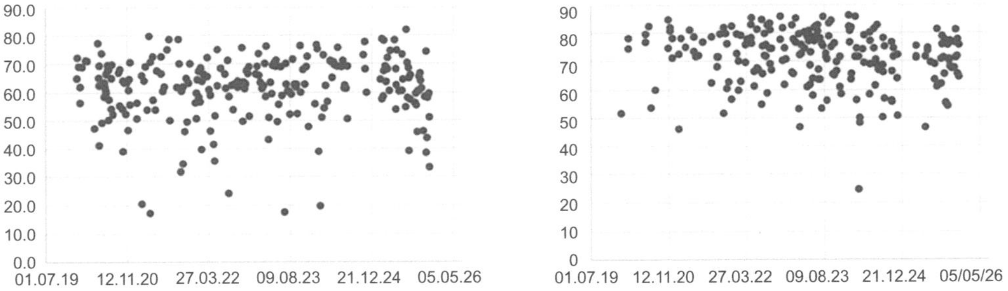

Mean Age	61,3 ys.
Parkinson	63,4 ys
Tremores	62,8 ys
Dystonia	55,0 ys

Mean Age	71,1 ys.
Parkinson	70,4 ys.
Tremores	71,9 ys.

Figure 10.
Age distribution of DBS and MRgFUS patients: even when not considering dystonia patients, DBS patients are nearly 10 years younger than MRgFUS patients.

Cost analyses indicate that DBS is associated with substantial initial expenditures [45, 46]. These include the costs of the surgical procedure, the implanted materials – such as electrodes and the IPG – as well as hospitalization costs, often including one night of postoperative monitoring in an intensive care unit. In addition, long-term follow-up costs must be considered, arising from outpatient visits for adjustment of stimulation parameters, routine follow-up assessments, and replacement of the pulse generator at intervals of approximately three to five years, depending on stimulation intensity [23]. This latter cost component has become less significant in recent years, as most patients now receive a rechargeable IPG. In this case, however, the initial costs of the pulse generator are higher.

Due to follow-up care and the configuration of the hardware, treatment may – depending on the patient population – entail ongoing involvement of relatives or nursing staff in order to ensure optimal outcomes. Overall, it becomes clear that DBS requires a specialized neurosurgical center with a multidisciplinary team, including dedicated neurorehabilitation clinics.

For patients with lower household incomes, travel costs may pose a significant barrier. For those not in salaried employment and without pension funds, DBS may result in a loss of income over several weeks. While DBS can significantly improve motor functions to the extent that patients are able to perform their work much more effectively in the long term, the procedure initially leads to a period of incapacity. Additionally, necessary follow-up visits for clinical monitoring and adjustment of stimulation parameters make temporary work absences unavoidable. Considering potential postoperative complications – such as infections requiring repeat surgical intervention – further periods of work incapacity may arise.

From a socio-economic perspective, it is evident that access to treatment with DBS is influenced by household income, socio-demographic background, living situation (rural vs. urban), and the type of health insurance coverage. Consequently, patients with higher household incomes, residing in urban areas, and covered by private health insurance generally gain earlier access to DBS therapy [47, 48].

In addition, it has become evident that – independent of the underlying disease and particularly with regard to movement disorders – male patients consult medical specialists at an earlier stage. Beyond the differences outlined above, this indicates the presence of a gender-specific inequality [49, 50]. This inequality affects both the timing of treatment and whether treatment with DBS is considered at all.

In summary, when evaluated over the long term, DBS is cost-effective compared with pharmacological therapy alone [21, 46]. However, it becomes apparent that a "triple bias" exists, as access to treatment is preferentially granted to affluent, urban-dwelling male patients.

Compared with DBS, the per-treatment costs of MRgFUS are lower, excluding the substantial initial investment required to procure the device [51]. If the device is not provided by regional health authorities, it can take several years for a hospital to recoup the procurement costs. MRgFUS does not involve the implantation of foreign material, thereby eliminating associated costs. Follow-up visits for stimulation parameter adjustments, as required in DBS, are also unnecessary since the procedure is typically completed in a single session. Each treatment requires a disposable treatment kit, which includes the phantom for daily quality assessment, the frame screws, and the membrane. Hospital stay costs are minimized, as patients are generally discharged the day after treatment, if not performed as an outpatient procedure. Most patients do not require subsequent inpatient rehabilitation, and long-term follow-up consists primarily of outpatient visits.

Overall, MRgFUS represents a cost-effective treatment option for movement disorders in eligible patients, particularly older patients with unilateral symptom predominance. Geographically, access is limited to specialized centers. In Germany, besides our center in Kiel, MRgFUS is currently offered in Bonn, Kassel, Duisburg, and, since 2025, Freiburg. In Kiel, after having started with the Exablate HiFUS in 2020, since 2025 we have utilized the latest "Exablate Prime" system (Insightec, Israel). The new system has the advantage that the 3-T-MRI scan can be run from the Prime Console, so no additional MRI technician is required for routine back-up.

Compared with DBS, MRgFUS may benefit both patients and health insurers through shorter hospital stays and faster recovery, allowing patients to return to work more quickly, provided they are still employed. Furthermore, permanent involvement of caregivers or nursing staff is generally unnecessary, avoiding additional indirect costs or financial burdens on families. Such help may be required in cases of DBS for assistance in recharging the battery or adjusting the settings if indicated.

From a long-term perspective, MRgFUS is more cost-efficient than DBS [51]. However, access is restricted to major centers, and natural anatomical limitations, such as SDR, can restrict eligibility – limitations that do not apply to DBS. Both MRgFUS and DBS disproportionately benefit urban, higher-income patients.

In Germany, statutory health insurance covers the costs of MRgFUS for patients with essential tremor who are not eligible for DBS, following a resolution by the Federal Joint Committee (G–BA) in 2021, with amendments in 2022. Private health insurers generally reimburse MRgFUS on a case-by-case basis ("individual case decision") upon submission of appropriate medical justification.

Ethical considerations in prioritizing treatment should weigh the long-term dependency on implanted devices and electronic equipment in DBS against the irreversibility of MRgFUS lesions, which cannot be adjusted once performed. Future efforts should address inequities in access to ensure more equitable treatment availability. Globally, both procedures will likely remain largely accessible only in countries with robust infrastructure and stable, publicly supported healthcare systems.

In summary, DBS offers flexibility and long-term symptom control but may deepen socioeconomic disparities, whereas MRgFUS is cost-efficient and patient-friendly, but access remains elite-centered.

5. Future directions in DBS and MRgFUS

DBS is undergoing a major transition from continuous, open-loop stimulation toward adaptive, patient-specific neuromodulation. Traditional DBS systems deliver constant stimulation independent of symptom fluctuations, whereas next-generation adaptive or closed-loop DBS systems aim to monitor neural or physiological signals in real time and adjust stimulation accordingly. These systems are already in advanced stages of development and have demonstrated the potential to improve clinical efficacy while reducing stimulation-related side effects and energy consumption [41].

Advances in computational modeling, machine learning, and artificial intelligence are central to this evolution. Model-based optimization algorithms, such as TuneS, are emerging to personalize electrode configurations and stimulation parameters, enabling improved engagement of disease-relevant neural circuits and potentially reducing adverse effects [42]. Closed-loop DBS platforms increasingly integrate sensing technologies and machine learning to decode motor or disease states and dynamically modulate stimulation in response, thereby supporting individualized neuromodulation strategies [43]. These developments may also address challenges related to stimulation consumption, habituation, and long-term efficacy.

In parallel with technological innovation, the clinical indications for DBS continue to expand. While Parkinson's disease and essential tremor remain core applications, DBS is being performed and/or investigated for a growing range of neurological and psychiatric conditions, including dystonia, treatment-resistant obsessive–compulsive disorder, major depressive disorder, Tourette's syndrome, chronic pain syndromes, and disorders of consciousness. These emerging indications highlight the need for refined target selection and stimulation paradigms, particularly for circuits involved in mood, cognition, and pain processing. In the more distant future, DBS is expected to evolve toward fully closed-loop, AI-driven systems capable of precise, circuit-level modulation across motor, affective, and cognitive domains.

MRgFUS is simultaneously advancing as a noninvasive neuromodulation and lesioning technology. Currently established for the treatment of essential tremor and tremor-dominant Parkinson's disease with favorable midterm outcomes, MRgFUS is increasingly being explored for additional indications such as neuropathic pain and psychiatric disorders. Ongoing technological improvements – including enhanced MRI thermometry, diffusion-based fiber tracking, and refined acoustic targeting – aim to improve targeting accuracy, expand patient eligibility, and minimize adverse effects [30].

Beyond its ablative applications, MRgFUS is emerging as a platform for therapeutic delivery. Growing evidence suggests that focused ultrasound can transiently and reversibly open the blood–brain barrier, facilitating the targeted delivery of pharmacological agents, biologics, or stem cells to specific brain regions [45–47]. This capability may significantly broaden the potential role of MRgFUS, extending its use beyond lesioning toward disease-modifying and restorative therapies.

Together, DBS and MRgFUS represent two complementary directions in modern neuromodulation: implantable systems offering precision tuning through adaptive, AI-assisted control, and incision-free, image-guided approaches enabling targeted intervention without implanted hardware. The future of functional neurosurgery will likely integrate both modalities, with individualized selection based on disease characteristics, therapeutic goals, and patient needs [44].

Ethical considerations always need to be discussed publicly and, in particular, with the patient and his family, as outcomes in general – and particularly changes in personality – are not always predictable, even with the most sophisticated systems. This is particularly relevant for DBS in disorders of consciousness, but also with psychiatric diseases.

6. Conclusion

DBS is evolving toward smarter, adaptive, and personalized therapy with closed-loop control and AI support, and is expanding into new clinical areas. MRgFUS is advancing beyond lesioning to broader neurological applications and potential therapeutic-delivery approaches, with ongoing improvements in safety and targeting. Both remain active fields of research, and their future lies in enhanced precision, carefully evaluated expansion of indications, and integration into multimodal neuromodulation paradigms.

Abbreviations

aDBS	adaptive Deep Brain Stimulation
cCT	Cranial Computed Tomography
DBS	Deep Brain Stimulation
ET	Essential Tremor
FDA	Food and Drug Administration
GPi	Globus Pallidus Internus
ICU	Intensive Care Unit
IPG	Implantable Pulse Generator
MRgFUS	Magnetic Resonance–Guided Focused Ultrasound
MRI	Magnetic Resonance Imaging
N	Nucleus (e.g. subthalamicus)
OCD	Obsessive–Compulsive Disorder
PD	Parkinson's Disease
STN	Subthalamic Nucleus
VIM	Ventral Intermediate Nucleus of the Thalamus

Acknowledgments

We thank our Department of Neurology, the referring physicians, and, above all, our patients for their continued cooperation.

Conflict of Interest

The authors declare having received travel grants from Medtronic, Boston Scientific, and InsighTec to attend meetings, and some received speakers' honoraria from Medtronic.

Author details

Bea Drucks, Hubertus Maximilian Mehdorn* and Isabel Luebbing
Dept. of Neurosurgery, University Clinics Schleswig-Holstein, Campus Kiel, Kiel, Germany

*Address all correspondence to: mehdorn@nch.uni-kiel.de

References

[1] Goetz CG. The history of Parkinson's disease: Early clinical descriptions and neurological therapies. Cold Spring Harbor Perspectives in Medicine. 2011;**1**(1):a008862

[2] Greenfield JG, Bosanquet FD. The brain-stem lesions in Parkinsonism. Journal of Neurology, Neurosurgery, and Psychiatry. 1953;**16**(4):213–226

[3] Hoehn MM, Yahr MD. Parkinsonism: Onset, progression and mortality. Neurology. 1967;**17**(5):427–442

[4] Spiegel EA, Wycis HT. Thalamotomy and pallidotomy for treatment of choreic movements. Acta Neurochirurgica (Wien). 1952;**2**(3–4): 417–22

[5] Al-rodhan NR, Kelly PJ. Pioneers of stereotactic neurosurgery. Stereotactic and Functional Neurosurgery. 1992;**58**(1–4):60–6

[6] Gildenberg PL. The birth of stereotactic surgery: A personal retrospective. Neurosurgery. 2004;**54**(1):199–207

[7] Spiegel EWHT. Stereoence phalotomy part I methods and stereotactic atlas of the human brain, 1952

[8] Spiegel EA, Wycis HT, Baird HW. Studies in Stereoencephalotomy. I. Topical relationships of subcortical structures to the posterior commissure. Confinia Neurologica. 1952;**12**(3):121–133

[9] Spiegel EA, Wycis HT, Freed H. Stereoencephalotomy in thalamotomy and related procedures. Journal of the American Medical Association. 1952;**148**(6):446–451

[10] Schaltenbrand GW. Atlas for Stereotaxy of the Human. Brain: Thieme, Stuttgart; 1977

[11] Nowinski WLF, Nguyen A, T B, Raphel JK, Jagannathan L, Raghavan R, Miller GA, et al. Multiple brain atlas database and atlas-based neuroimaging system. Computer Aided Surgery 1997;**2**:42–66

[12] Roth A, Buttrick SS, Cajigas I, Jagid JR, Ivan ME. Accuracy of frame-based and frameless systems for deep brain stimulation: A meta-analysis. Journal of Clinical Neuroscience. 2018;**57**:1–5

[13] Benabid AL. Deep brain stimulation for Parkinson's disease. Current Opinion in Neurobiology. 2003;**13**(6):696–706

[14] Wichmann T, DeLong MR, Guridi J, Obeso JA. Milestones in research on the pathophysiology of Parkinson's disease. Movement Disorders. 2011;**26**(6):1032–1041

[15] Benabid AL, Chabardes S, Mitrofanis J, Pollak P. Deep brain stimulation of the subthalamic nucleus for the treatment of Parkinson's disease. The Lancet Neurology. 2009;**8**(1):67–81

[16] Benabid AL, Pollak P, Gervason C, Hoffmann D, Gao DM, Hommel M, et al. Long-term suppression of tremor by chronic stimulation of the ventral intermediate thalamic nucleus. Lancet. 1991;**337**(8738):403–6

[17] Limousin P, Krack P, Pollak P, Benazzouz A, Ardouin C, Hoffmann D, et al. Electrical stimulation of the subthalamic nucleus in advanced Parkinson's disease. New England Journal of Medicine 1998;**339**:1105–1111

[18] Pinsker MO, Herzog J, Falk D, Volkmann J, Deuschl G, Mehdorn M. Accuracy and distortion of deep brain stimulation electrodes on postoperative MRI and CT. Zentralbl Neurochir. 2008;**69**(3):144–147

[19] Pinsker MO, Volkmann J, Falk D, Herzog J, Alfke K, Steigerwald F, et al. Electrode implantation for deep brain stimulation in dystonia: A fast spin-echo inversion-recovery sequence technique for direct stereotactic targeting of the GPI. Zentralbl Neurochir 2008;**69**:71–75

[20] The Noble Art of Lesioning Marseille. 2025

[21] Dams J, Siebert U, Bornschein B, Volkmann J, Deuschl G, Oertel WH, et al. Cost-effectiveness of deep brain stimulation in patients with Parkinson's disease. Movement Disorders 2013;**28**:763–771

[22] Deuschl G, Schade-Brittinger C, Krack P, Volkmann J, Schafer H, Botzel K, et al. A randomized trial of deep-brain stimulation for Parkinson's disease. New England Journal of Medicine 2006;**355**:896–908

[23] Lhommee E, Wojtecki L, Czernecki V, Witt K, Maier F, Tonder L, et al. Behavioural outcomes of subthalamic stimulation and medical therapy versus medical therapy alone for Parkinson's disease with early motor complications (EARLYSTIM trial): Secondary analysis of an open-label randomised trial. The Lancet Neurology 2018;**17**:223–231

[24] Schuepbach WMM, Tonder L, Schnitzler A, Krack P, Rau J, Hartmann A, et al. Quality of life predicts outcome of deep brain stimulation in early Parkinson disease. Neurology 2019;**92**: e1109–e20

[25] Krause P, Mahlknecht P, Skogseid IM, Steigerwald F, Deuschl G, Erasmi R, et al. Long-term outcomes on pallidal neurostimulation for dystonia: A controlled, prospective 10-year follow-up. Movement Disorders 2025;**40**:1098–1111

[26] Hamel W, Koppen JA, Alesch F, Antonini A, Barcia JA, Bergman H, et al Targeting of the subthalamic nucleus for deep brain stimulation: A survey among parkinson disease specialists. World Neurosurgery. 2017;**99**:41–46

[27] McIntyre CC, Savasta M, Walter BL, Vitek JL. How does deep brain stimulation work? Present understanding and future questions. Journal of Clinical Neurophysiology. 2004;**21**(1):40–50

[28] Kuhn AA, Kempf F, Brucke C, Gaynor Doyle L, Martinez-Torres I, Pogosyan A, et al. High-frequency stimulation of the subthalamic nucleus suppresses oscillatory beta activity in patients with Parkinson's disease in parallel with improvement in motor performance. Journal of Neuroscience 2008;**28**:6165–6173

[29] McIntyre CC, Savasta M, Kerkerian-Le Goff L, Vitek JL. Uncovering the mechanism(s) of action of deep brain stimulation: Activation, inhibition, or both. Clinical Neurophysiology. 2004;**115** (6):1239–1248

[30] Bekar L, Libionka W, Tian GF, Xu Q, Torres A, Wang X, et al. Adenosine is crucial for deep brain stimulation-

mediated attenuation of tremor. Nature Medicine 2008;**14**:75–80

[31] Bikovsky L, Hadar R, Soto-Montenegro ML, Klein J, Weiner I, Desco M, et al. Deep brain stimulation improves behavior and modulates neural circuits in a rodent model of schizophrenia. Experimental Neurology. 2016;**283**(Pt A):142–50

[32] Johnson KA, Worbe Y, Foote KD, Butson CR, Gunduz A, Okun MS. Tourette syndrome: Clinical features, pathophysiology, and treatment. The Lancet Neurology. 2023;**22**(2):147–158

[33] Meyer GM, Hollunder B, Li N, Butenko K, Dembek TA, Hart L, et al. Deep brain stimulation for obsessive-compulsive disorder: Optimal stimulation sites. Biological Psychiatry 2024;**96**:101–113

[34] Neudorfer C, Kultas-Ilinsky K, Ilinsky I, Paschen S, Helmers AK, Cosgrove GR, et al. The role of the motor thalamus in deep brain stimulation for essential tremor. Neurotherapeutics 2024;**21**:e00313

[35] Ryvlin P, Rheims S, Hirsch LJ, Sokolov A, Jehi L. Neuromodulation in epilepsy: State-of-the-art approved therapies. The Lancet Neurology. 2021;**20**(12):1038–1047

[36] Whiting AC, Oh MY, Whiting DM. Deep brain stimulation for appetite disorders: A review. Neurosurgical Focus. 2018;**45**(2):E9

[37] Yu D, Yan H, Zhou J, Yang X, Lu Y, Han Y. A circuit view of deep brain stimulation in Alzheimer's disease and the possible mechanisms. Molecular Neurodegeneration. 2019;**14**(1):33

[38] Cheyuo C, Vetkas A, Sarica C, Kalia SK, Hodaie M, Lozano AM. Comprehensive characterization of intracranial hemorrhage in deep brain stimulation: A systematic review of literature from 1987 to 2023. Journal of Neurosurgery. 2024;**141**(2):381–393

[39] Elias WJ, Lipsman N, Ondo WG, Ghanouni P, Kim YG, Lee W, et al. A randomized trial of focused ultrasound thalamotomy for essential tremor. New England Journal of Medicine 2016;**375**:730–739

[40] Cosgrove GR, Lipsman N, Lozano AM, Chang JW, Halpern C, Ghanouni P, et al. Magnetic resonance imaging-guided focused ultrasound thalamotomy for essential tremor: 5-year follow-up results. Journal of Neurosurgery 2023;**138**:1028–1033

[41] Armengou-Garcia L, Sanchez-Catasus CA, Aviles-Olmos I, Jimenez-Huete A, Montoya-Murillo G, Gorospe A, et al. Unilateral magnetic resonance-guided focused ultrasound lesion of the subthalamic nucleus in Parkinson's disease: A prospective study. Movement Disorders 2024;**39**:2230–2241

[42] Rodriguez-Rojas R, Pineda-Pardo JA, Martinez-Fernandez R, Kogan RV, Sanchez-Catasus CA, Del Alamo M, et al. Functional impact of subthalamotomy by magnetic resonance-guided focused ultrasound in Parkinson's disease: A hybrid PET/MR study of resting-state brain metabolism. European Journal of Nuclear Medicine and Molecular Imaging 2020;**47**:425–436

[43] Abbas A, Sabet H, Abouelmagd ME, Hefnawy MT, Bakr S, Lashin B, et al. Magnetic resonance-guided focused ultrasound subthalamotomy for Parkinson's disease: A meta-analysis of

effectiveness and safety. Neurosurgical Review 2026;**49**:201

[44] Boutet A, Gwun D, Gramer R, Ranjan M, Elias GJB, Tilden D, et al. The relevance of skull density ratio in selecting candidates for transcranial MR-guided focused ultrasound. Journal of Neurosurgery 2020;**132**:1785–1791

[45] Dang TTH, Rowell D, Connelly LB. Cost-effectiveness of deep brain stimulation with movement disorders: A systematic review. Movement Disorders Clinical Practice. 2019;**6**(5):348–358

[46] Sasidharan A, Bagepally BS, Kumar SS. Cost effectiveness of deep brain stimulation for Parkinson's disease: A systematic review. Applied Health Economics and Health Policy. 2024;**22**(2):181–192

[47] Porst M, von der Lippe E, Leddin J, Anton A, Wengler A, Breitkreuz J, et al. The Burden of Disease in Germany at the National and Regional Level. Dtsch Arztebl Int. 2022;**119**(46):785–92

[48] Cavaleri J, Stefanescu K, Lee D, Mason X. Deep brain stimulation access in 2025: Geographic, gender, racial, and socioeconomic disparities re-examined. Current Opinion in Neurology. 2025;**38**(4):330–336

[49] Hendriks M, Vinke RS, Georgiev D. Gender discrepancies and differences in motor and non-motor symptoms, cognition, and psychological outcomes in the treatment of Parkinson's disease with subthalamic deep brain stimulation. Frontiers in Neurology. 2023;**14**:1257781

[50] Jost ST, Strobel L, Rizos A, Loehrer PA, Ashkan K, Evans J, et al. Gender gap in deep brain stimulation for Parkinson's disease. NPJ Parkinson's Disease 2022;**8**:47

[51] Jameel A, Meiwald A, Bain P, Patel N, Nandi D, Jones B, et al. The cost-effectiveness of unilateral magnetic resonance-guided focused ultrasound in comparison with unilateral deep brain stimulation for the treatment of medically refractory essential tremor in England. British Journal of Radiology. 2022;**95**:1140

Section 3

Technology, Systems, and the Future of Neurosurgical Practice

Chapter 9

Patient Positioning During Neurosurgical Procedures

Ilshat Gaysin and Egor Markin

Abstract

This chapter delves into the critical role of patient positioning in neurosurgery, highlighting its significance in ensuring both safety and optimal surgical outcomes. It begins by outlining the various positioning techniques used in different neurosurgical procedures, such as supine, prone, and lateral positions, and discusses the anatomical considerations that influence these choices. The chapter then examines the consequences of inadequate positioning. It emphasizes the need for thorough preoperative assessments and the implementation of comprehensive positioning protocols. By combining foundational principles with innovative approaches, this chapter aims to empower neurosurgeons and their teams to optimize patient positioning and enhance surgical efficacy in diverse clinical scenarios.

Keywords: head fixation, operating table, intracranial pressure, complication, patient position

1. Introduction

Achieving correct patient positioning constitutes a pivotal initial phase in neurosurgical operations, directly influencing procedural efficacy and outcomes. Strategic patient positioning is a critical preoperative determinant of success. It directly enhances surgical workflow efficiency while systematically decreasing the risk of perioperative complications.

2. Organization of the operating room during neurosurgical procedures

Every neurosurgical procedure is a team effort, whose success relies on each member. Optimizing this work begins with the operating room setup – strategically placing equipment and arranging the table to ensure an efficient and ergonomic workflow.

An efficient operating room setup ensures unobstructed access to equipment for every team member. For the anesthesiologist, this means constant, ready access to intravenous catheters, endotracheal tubes, and monitoring sensors. This access is often facilitated by using a specialized operating table or a fixed metal arch, which creates a protected space under the sterile drapes for necessary manipulation. The scrub nurse must be positioned for optimal efficiency, with direct access to the main sterile instrument table. This setup enables the rapid retrieval of additional tools, which is critical during urgent phases of the operation. For optimal ergonomics with a right-handed surgeon, the scrub nurse is typically stationed to the surgeon's right to facilitate smooth instrument exchange.

Basic operating room organization schemes are presented in **Figure 1**. Complex procedures utilizing awake surgery, intraoperative neuromonitoring, or intraoperative neuroimaging often necessitate adjustments to the surgical layout, including equipment placement and even patient positioning.

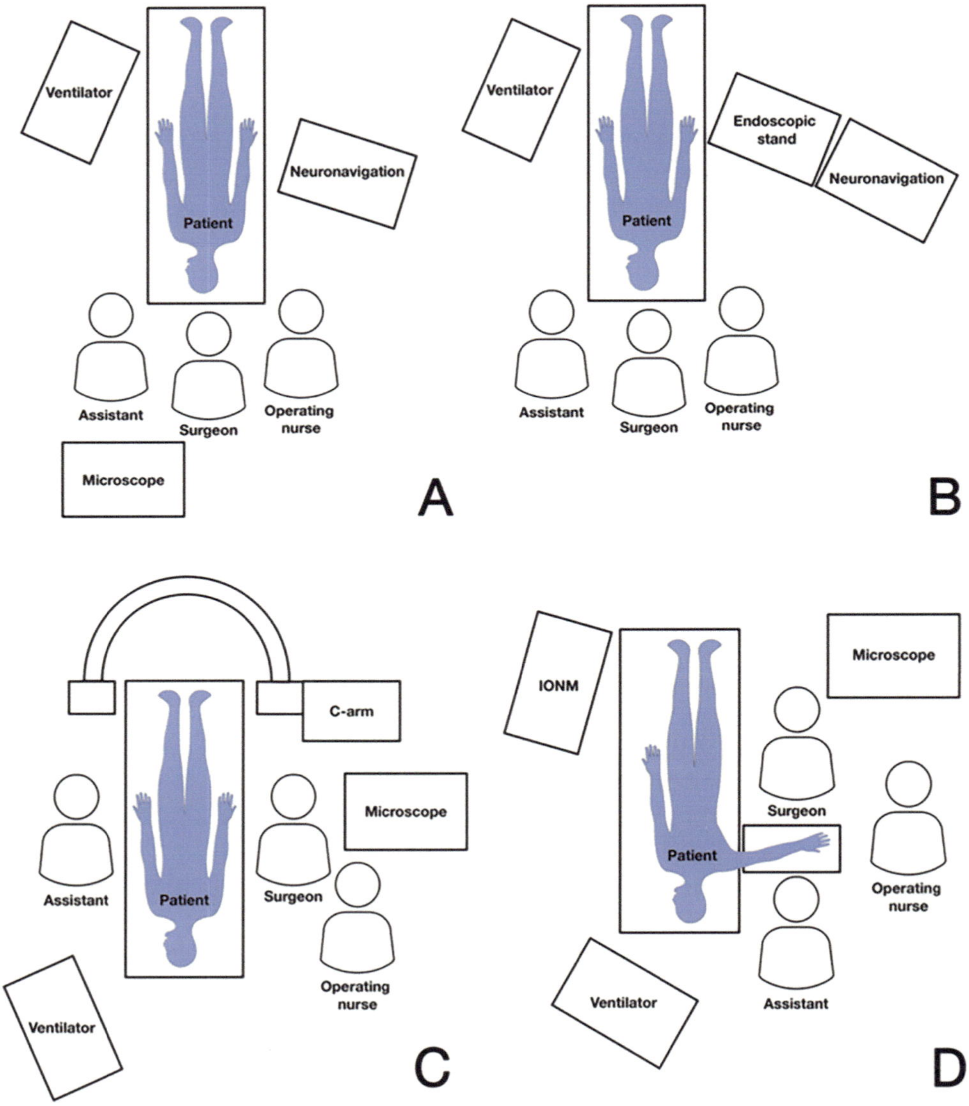

Figure 1.
Basic operating room organization schemes for neurosurgical operations in conventional (A) cranial, (B) neuroendoscopic, (C) spinal procedures, (D) peripheral nerve surgeries.

3. Main principles of patient positioning

The head connects to the body through the mobile neck – a complex structure of vertebrae, muscles, neurovascular structures, and soft tissues. This anatomy provides wide mobility with three primary movements: flexion/extension, lateral flexion (right and left), and rotation. While the body also offers significant positional flexibility, the final surgical position must be determined through coordinated teamwork and strict adherence to the main principles that ensure patient safety and maintain operative success.

Head Fixation. The millimeter-to-submillimeter precision required in modern neurosurgery, particularly in ultramicroneurosurgery, is fundamentally dependent on rigid cranial fixation. This immobilization is most reliably achieved using specialized skull clamps, primarily the three to four-pin Mayfield (**Figure 2**) and the four to six-pin Sugita. Despite their distinct designs, both fixation systems carry a comparable complication risk of approximately 1%, including skull fracture, epidural hematoma, and neurovascular injury [1].

The safe and stable application of the Mayfield clamp is governed by several key principles. First, pin placement must meticulously avoid critical structures, including neurovascular trunks, thin areas of cranial bone (e.g., the temporal squama or areas overlying dural sinuses), sites of prior craniotomy or fracture, and the muscular bulk of the temporalis and trapezius. Second, optimal stability is achieved by positioning two pins on the inferior half of the skull relative to its equator, in accordance with the "soccer ball concept." This also involves orienting the clamp's central axis below the cranial midline to secure the skull and minimize the risk of slippage. Finally, the area of latitude inferior to the temporal line provides an anatomical landmark for safe pin placement [2].

For procedures not requiring rigid fixation – including many spinal surgeries – a horseshoe headrest offers effective soft immobilization. These headrests are available in distinct adult and pediatric sizes to accommodate anatomical differences. When positioning a patient in the prone, semi-prone, or Concord position using a horseshoe headrest or silicone pads for head support, meticulous attention must be given to eye protection. Prolonged or misplaced pressure on the orbits can lead to conjunctival edema and, in severe cases, vision loss [3].

Pediatric head fixation presents unique anatomical challenges, including open fontanels, thin and mobile cranial bones, and ongoing ossification. Consequently, no

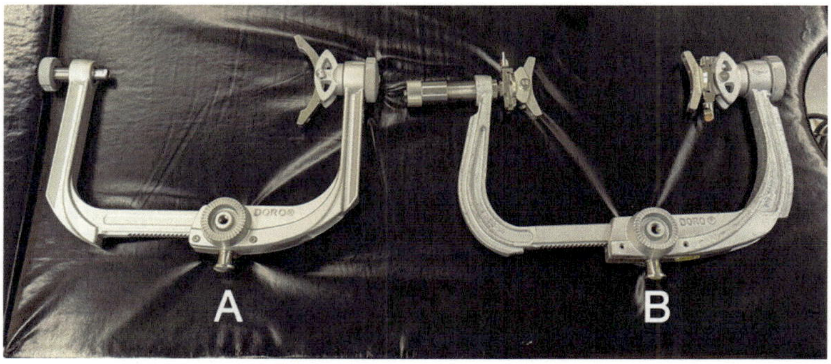

Figure 2.
Design modification options for (A) 3-point and (B) 4-point fixation of the Doro/Mayfield skull clamps.

single fixation method is universally standardized, though the Mayfield skull clamp for rigid fixation is the most frequently documented system in the literature. For pediatric patients, Mayfield clamp placement and pin fixation should be used with extreme caution under the age of 2, with progressively increasing pressures applied: 10-20 lbs (2-3 years) and 21-30 lbs (3–4 years) using pediatric pins, then 21-40 lbs (4–5 years) with pediatric pins, 31-50 lbs (5-10 years) with pediatric or adult pins, and finally 60 lbs with adult pins for children over 10 (**Figure 3**) [4].

Several alternative fixation methods are also described, including pinning with a modified rubber stopper technique, hybrid approaches that combine a Mayfield clamp with a horseshoe headrest, and adaptations using a sterile U-drape with a horseshoe headrest, a Doro clamp (modified Mayfield clamp) in combination with silicone pads instead of pins, and other techniques, as presented in **Figure 4** [5–7].

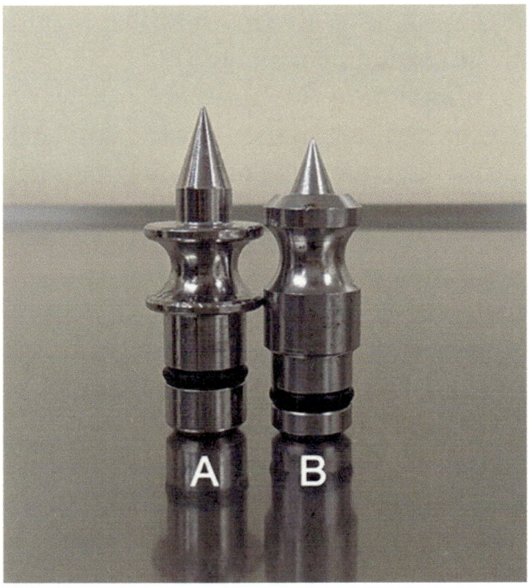

Figure 3.
Different designs of (A) adult and (B) pediatric pins.

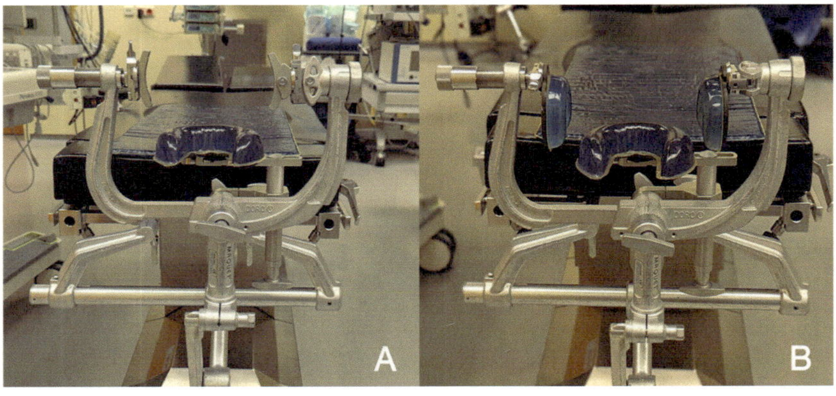

Figure 4.
Illustration of options for fixing a child's head using a combination of (A) a horseshoe headrest and a 4-point Doro clamp and (B) three horseshoe-shaped silicone pads.

Positioning tools. Effective positioning devices must fulfill several key biomechanical and physiological functions: they must absorb compressive forces, distribute pressure evenly to prevent focal ischemia of surface tissue, protect against excessive tissue stretch or compression, and permit adequate thoracic expansion for ventilation and gas exchange [8].

To mitigate pressure on vulnerable bony prominences during prolonged neurosurgical procedures, the use of specialized soft padding – most commonly silicone pads – is essential. The specific anatomical areas at risk vary with the patient's position and include the scapulae, elbows, sacrum, ischial tuberosities, iliac crests, greater trochanters, knees, and heels. Meticulous attention to this aspect of positioning is a critical preventive measure against the development of pressure sores and other compression-related injuries (**Figure 5**).

In the three-quarter prone, or park bench position, specific attention must be paid to limb positioning. The dependent arm should be comfortably supported to avoid excessive traction; the use of compression bandages on this limb can help prevent dependent edema and reduce the risk of thrombosis. For the lower limbs, a soft pad must be placed between them. The uppermost or lower leg should be flexed at the hip and knee to enhance the patient's stability in the final position. To maintain a stable body position, it is necessary to use lateral supports attached to the operating table.

When performing spinal procedures in the prone or knee-chest position, the use of a Wilson frame or supportive pads is essential to create free space in the abdominal area. This reduces intra-abdominal pressure, which, if elevated, increases spinal venous pressure and consequently leads to greater surgical bleeding.

The most convenient method is to use a special vacuum mattress, which, after the patient is positioned and negative pressure is created, adapts to the shape of the adjacent body surface, evenly distributing the load.

"Gravity is your friend". Head positioning requires precise planning of the surgical trajectory through the relevant anatomical corridors, ensuring it is oriented toward the surgeon in an angled view (**Figure 6**). This strategic alignment utilizes gravity as a "natural retractor", which significantly improves intraoperative

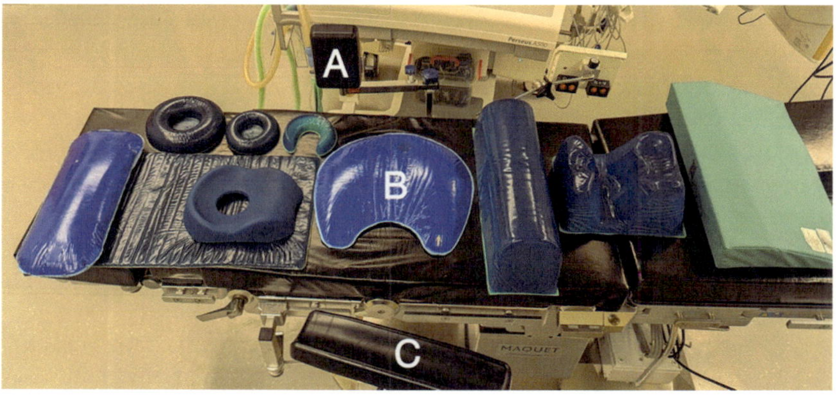

Figure 5.
Various instruments for patient positioning (A) lateral Support, (B) silicone and soft pads of various shapes, (C) multi-position arm board.

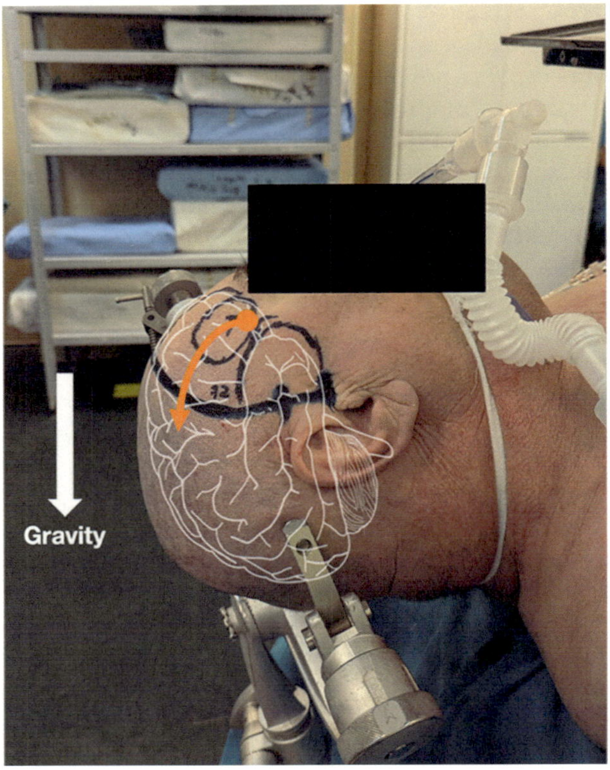

Figure 6.
Schematic illustration demonstrat on of the use of gravity (White Arrow) assistance, for retraction for surgical access to the tumor beneath the frontal lobe (orange arrow). Head Fixed using the Doro/Mayfield skull clamp.

visualization and minimizes the need for manual brain retraction, thereby enhancing patient safety.

The strategic use of gravity aids in the gentle separation of the brain from the skull base and in the dissection of cerebral fissures. This principle is effectively employed in most key approaches. For instance, in the supracerebellar approaches with the patient in a sitting position, gravity retracts the cerebellum. Similarly, for an interhemispheric approach, rotating the head to 90 degrees allows gravity to assist in hemisphere retraction, providing access to both ipsilateral and contralateral parasagittal lesions.

Patient position and physiological responses. The patient's position for surgery directly influences the physiological response. Changes in body and head positioning alter cerebrospinal fluid (CSF) dynamics and cerebral hemodynamics. Through the Monro-Kellie doctrine, these alterations directly impact intracranial pressure, which in certain situations can exacerbate pathological brain conditions and complicate surgical management.

Head elevation causes a partial displacement of CSF into the spinal canal. To mitigate the effects of this shift, positional changes should be performed slowly and deliberately. This reduces the volume, rate, and velocity of CSF flow, allowing time for safe redistribution. For patients at elevated risk, preoperative administration of mannitol and hyperventilation can be employed to increase intracranial compliance, thereby preparing the CSF space for postural changes [9].

Excessive rotation, flexion (two-finger breadth between the mandible and clavicle), or extension of the head risks obstructing venous outflow. Due to anatomical variability – most commonly a right-dominant cerebral venous sinus – this risk must be meticulously integrated into preoperative planning and intraoperative positioning to prevent neck vein compression. This is especially critical in overweight patients. Impaired venous drainage can elevate both venous pressure, increasing surgical bleeding, and intracranial pressure, a principle historically demonstrated by the Queckenstedt maneuver [10, 11].

Patient-specific conditions. For cranial procedures in the prone or Concord position, abdominal pressure management is particularly critical in obese patients. The use of specialized pads to offload the abdomen is essential to prevent a cascade of increased venous pressure, which aggravates surgical bleeding and can raise intracranial pressure. Excess cervical adipose tissue also increases the risk that excessive angulation will compress the jugular veins and disrupt venous drainage. This risk must be carefully assessed and mitigated during head fixation.

Prolonged surgical procedures significantly increase thromboembolic risk, particularly in high-risk patients. In addition to mandatory pharmacological prophylaxis, adjunct mechanical measures, such as intermittent pneumatic compression devices for the lower extremities, are recommended. These devices can be employed intraoperatively and continued postoperatively to enhance protective efficacy [12–14].

4. Patient positioning

The optimal patient position for a neurosurgical intervention is selected by balancing three priorities: uncompromised patient safety, optimal access to the pathology, and surgeon ergonomics. The surgical literature accordingly documents a wide variety of positioning techniques and specific modifications tailored to diverse neurological conditions. Guided by the main positioning principles, the patient can be placed in a variety of standard configurations and their institutional modifications. The most common positions in neurosurgery include the supine, lateral, park bench, three-quarter-prone, prone, sitting, and Concorde positions (**Figure** 7).

4.1 Supine position

As the most common and adaptable position in cranial neurosurgery, the supine position is employed for a wide variety of surgical approaches (**Figure 8**). The cervical spine's extensive range of motion enables optimal head positioning, providing surgical access to most brain areas. Placing a soft pad under the shoulder contralateral to the desired head rotation or tilting the table facilitates near-90-degree turns, which enables posterior fossa access via subtemporal or retrosigmoid approaches, or even possible access to the pineal region or brainstem [15, 16]. Throughout this process, careful attention must be paid to avoid jugular venous compression.

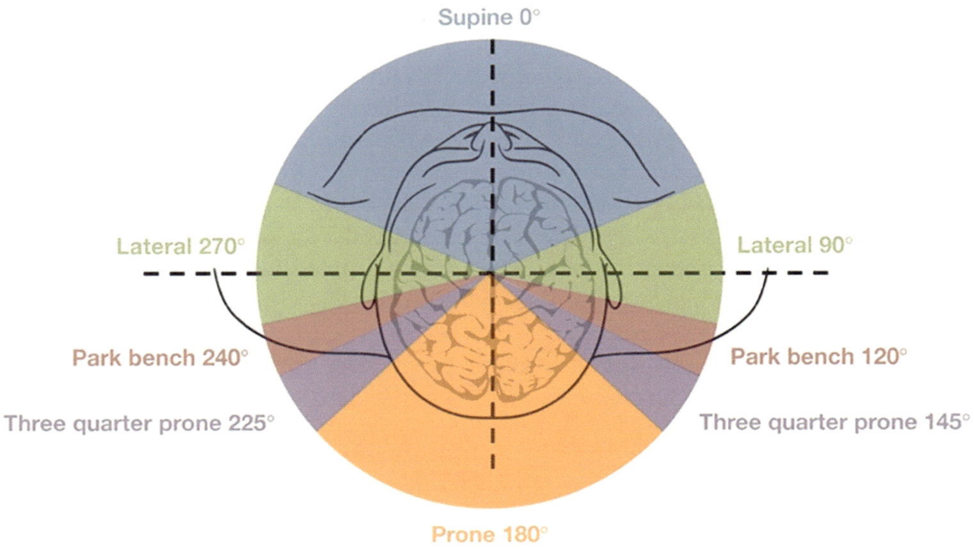

Figure 7.
Schematic representation of the angles of head rotation in different positions.

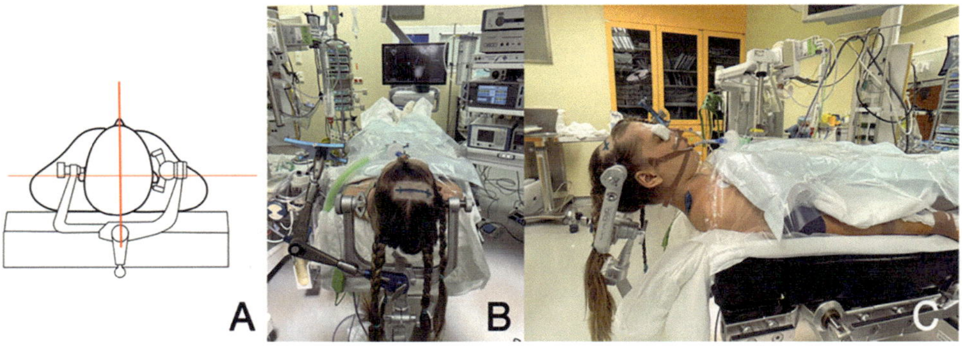

Figure 8.
(A) Supine position scheme, (B) intraoperative position of the patient, view from above, (C) side view.

Three variants of the supine position are utilized in neurosurgery. The standard horizontal supine position is poorly tolerated by conscious patients over extended periods and is therefore seldom recommended. In contrast, the lawn-chair position – with the head and thorax slightly elevated and the hips and knees flexed – represents a more physiological and tolerable alignment. This configuration promotes cerebral venous drainage and enhances systemic venous return. A third variant, the reverse Trendelenburg position, modifies the horizontal plane by elevating the head of the table [17].

The supine position is also indicated for anterior approaches to the spine, such as anterior lumbar interbody fusion, and for operations on the cervical vasculature and spine. Furthermore, some surgeons favor it over the semi-sitting position for endoscopic transnasal procedures.

4.2 Lateral position

The lateral position involves placing the patient on their side, with the head rotated approximately 90 degrees (**Figure 9**). This orientation provides optimal access to the lateral cerebral surface, anterior and middle cranial fossae, and is indicated for subtemporal, retrosigmoid, and other laterally-based approaches to the posterior fossa. Additionally, it effectively utilizes gravity to facilitate brain retraction in the interhemispheric approach [18].

4.3 Park bench position

A modification of the standard lateral position, the park bench position is designed to improve visualization and access within the posterior fossa. By extending the shoulder and arm beyond the table, it permits greater shoulder rotation and enhanced flexibility for positioning of the head [19]. The head is turned at an angle of about 120 degrees.

The park bench position is optimized by slightly elevating the head of the table while keeping the head horizontal, which improves the angle between the neck and shoulder for surgical access. Careful attention must be paid to the dependent arm, maintaining a neutral, physiological alignment by using a multi-position arm board. It is critical to pad the axilla adequately to prevent excessive compression and reduce the risk of brachial plexopathy. As with the standard lateral position, secure the patient using lateral table supports and apply soft padding to all prominent pressure points (**Figure 10**).

The modified park bench position is the Dormeuse position that incorporates key features of the semi-sitting position. In this position, the patient is positioned by elevating the head section of the operating table approximately 30°, ensuring the thorax is above the level of the feet. The hips and knees are gently flexed, with a supportive pillow placed between the knees. Slight leg elevation is achieved by adjusting the leg section of the table or by placing pillows beneath the calves. The head is maintained in strict axial alignment with the thorax, without lateral flexion. Finally, the head is rotated toward the side of the craniotomy and placed in slight anteflexion. By elevating the head above the heart, posterior fossa venous pressure is

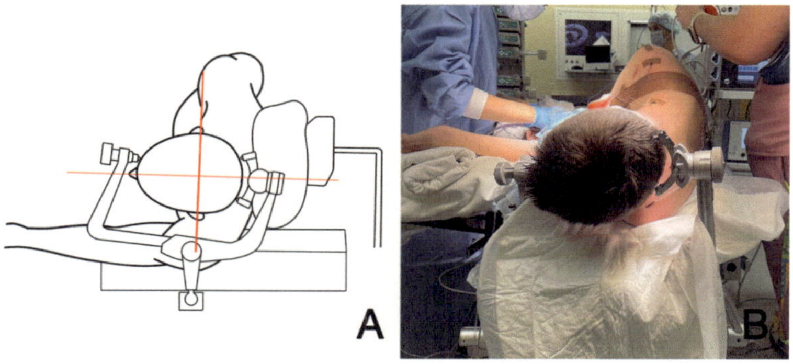

Figure 9.
(A) Lateral position scheme, (B) intraoperative position of the patient viewed from above.

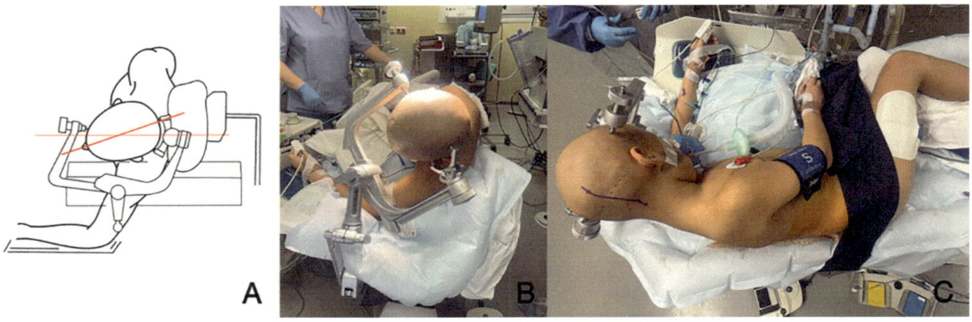

Figure 10.
(A) Park bench position scheme, (B) intraoperative position of the patient viewed from above, (C) side view.

reduced, and the cerebellum is allowed to retract from the petrous bone and tentorium [20].

4.4 Three quarter prone position

Similar to the park bench position, the three-quarter prone position involves a more pronounced rotation toward the prone plane, with the head angled at approximately 135 degrees (**Figure 11**). It is primarily indicated for surgical approaches to the posterior fossa, the parieto-occipital area, and the pineal region [21]. As with the park bench position, the head end of the table can be slightly elevated to increase the angle between the neck and shoulder, optimizing access to the craniovertebral junction.

4.5 Prone position

The prone position places the patient in a horizontal, face-down orientation. It is the standard position for surgical access to the spinal axis and is also employed for

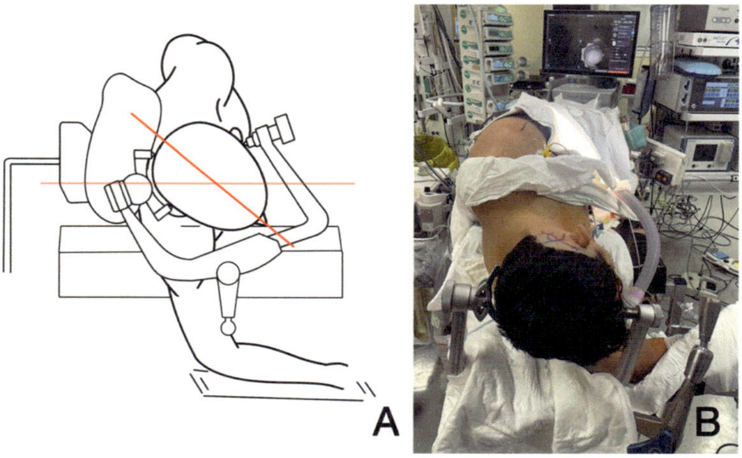

Figure 11.
(A) Three-quarter prone position scheme, (B) intraoperative position of the patient viewed from above.

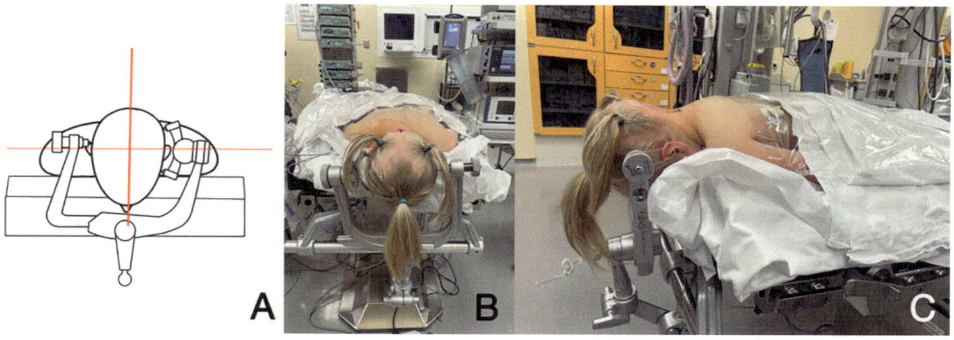

Figure 12.
(A) Prone position scheme, (B) intraoperative position of the patient viewed from above, (C) side view.

pathologies of the posterior cranial fossa, occipital and parieto-occipital regions, and the craniovertebral junction (**Figure 12**). The head is typically supported in a neutral or rotated alignment using a horseshoe headrest, a specialized silicone pad, or rigid fixation. The highest level of evidence regarding prone positioning focuses on two complications: postoperative vision loss and cardiovascular events, including hypovolemia and cardiac arrest [22].

A notable modification of the standard prone position is the "sphinx" or "sea-lion" position, first described by Suzuki et al. in 1983 [23]. It is primarily indicated for pediatric procedures such as total calvarial reconstruction and strip craniectomy, as well as for adult surgeries targeting falcotentorial lesions. The patient is placed prone, with the head supported on a horseshoe headrest at the malar eminences. To prevent cervical hyperextension, the table is placed at 30–45 degrees of reverse Trendelenburg. This configuration, combined with intraoperative table tilting, provides simultaneous access to both the anterior and posterior halves of the skull [23].

4.6 Sitting position

Positioning a patient in the sitting posture presents distinct anatomical and physiological challenges. The standard sitting (Fowler's) position is defined by a 45–60-degree trunk elevation, with knees either flexed or extended. A common variation is the semi-sitting position, characterized by less head elevation (30–45 degrees) and elevation of the foot of the bed to flex the knees and legs (**Figure 13**). A modified sitting position employs a steep Trendelenburg tilt with the patient's back elevated. The lower extremities are placed in semi-flexion on supportive pillows to minimize vascular kinking and reduce the risk of venous stasis [24]. The lounging position is another variation of the sitting position. It maintains the traditional sitting position but involves elevating the lower extremities to or slightly above the level of the head [25].

The sitting position is indicated for accessing the cerebellopontine angle, pineal region, occipital region, most structures in the posterior cranial fossa, and the cervical spine. Its advantages include improved drainage of CSF and blood from the surgical field, reduced venous bleeding, and gravitational retraction of the cerebellum. However, these benefits must be weighed against significant disadvantages and risks.

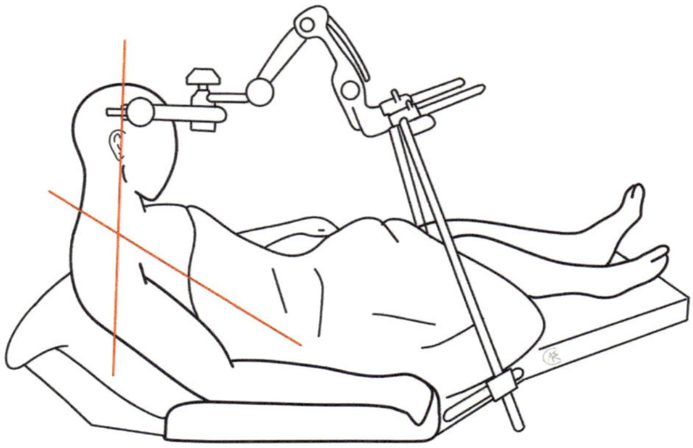

Figure 13.
Sitting position scheme illustration.

The sitting position presents significant risks during procedures, most critically venous air embolism and intraoperative hypotension. The reported incidence of venous air embolism is 39% for posterior fossa procedures and 12% for cervical spine surgeries, while intraoperative positional hypotension occurs in approximately 37% of cases [26, 27]. Venous air embolism occurs due to the elevated position of the head above the heart, generating negative pressure within the cranial veins. Performing surgery in this position, therefore, mandates an experienced team with established protocols for managing these complications.

Preoperative assessment requires an echocardiogram to exclude cardiac pathologies such as heart failure or an atrial septal defect, which increase the risk of paradoxical embolism. Intraoperatively, continuous monitoring via transesophageal echocardiography and end-tidal CO_2 ($EtCO_2$) is essential. The severity of any embolism is graded using the Tübingen classification system. Should a severe or uncontrolled venous air embolism occur, immediate intervention is necessary to increase venous pressure by tilting the table into Trendelenburg, lowering the head, or temporarily applying jugular compression, followed by the rapid elimination of the air entry source and/or aspirating the air through a central venous catheter.

4.7 Concorde position

The Concorde position was first introduced by Kobayashi et al. in 1983 as an alternative to the sitting position for approaching the pineal region. The patient is placed in a prone position, positioned as far left on the table as possible. The head is elevated above the level of the heart, then fixed in a posture of flexion and rightward tilt (**Figure 14**) [28]. The surgeon operates from the patient's left side.
A modification incorporates "skew head rotation," where the head is both tilted and rotated by adjusting the head frame intraoperatively to optimize the surgical trajectory [29]. While the original technique places the arms alongside the body, an "arm-down" variation lowers the ipsilateral arm to enhance visualization. This adjustment is particularly beneficial for muscular, broad-shouldered, short-necked, or obese patients [30]. The Concorde position offers optimal access to the pineal region and posterior cranial fossa – particularly the fourth ventricle – while significantly reducing the risk of venous air embolism compared to the sitting

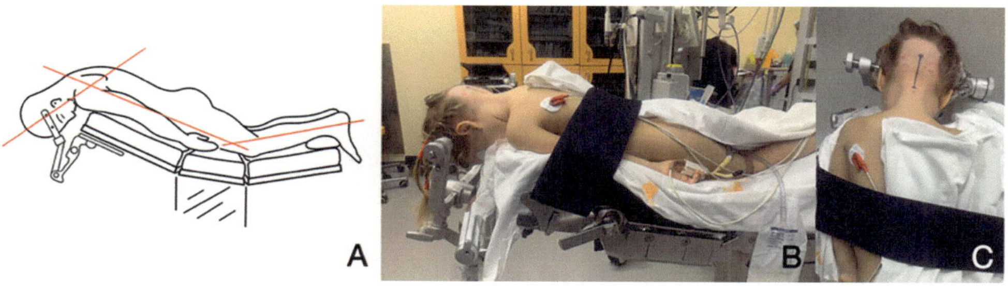

Figure 14.
(A) Concorde position scheme, (B) intraoperative position of the patient side view, (C) back view.

position. The literature lacks systematic studies on complications associated with the Concorde position; available reports are limited to isolated cases of lingual abscess, thoracic spinal cord injury, and brain death [31–33]. This issue is the subject of further research.

5. Conclusion

Patient positioning is a critical determinant of success in neurosurgery, requiring careful planning and precise execution. The optimal position balances two fundamental goals: providing the surgeon with optimal exposure to the pathology while ensuring the patient's safety. This decision-making process integrates anatomical knowledge, the selected approach, the main principles of positioning, and patient position variants. Ultimately, safe and effective positioning is a coordinated team effort, applying standardized principles while adapting to the unique demands of each patient and procedure to create the foundation for a successful surgical procedure.

Acknowledgment

The authors acknowledge the use of DeepSeek for language polishing of the manuscript.

Author details

Ilshat Gaysin[1]* and Egor Markin[2]

1 Regional Emergency Medical Center named after R.S. Akchurin, Naberezhnye Chelny, Russian Federation

2 Oncology and Immunology, Dmitry Rogachev National Research Center for Pediatric Hematology, Moscow, Russian Federation

*Address all correspondence to: gaysin.ilshat.a@gmail.com

References

[1] Al-Juboori AA, Badran SA, Qasim AM, Shahadha AA, Alsamok AS, Al-Taie RH, et al. Head Fixation devices in neurosurgery: A systematic review and meta-analysis. World Neurosurgery. 2025;**202**:124348

[2] Thijs D, Menovsky T. The Mayfield Skull Clamp: A literature review of its complications and technical nuances for application. World Neurosurgery. 2021;**151**:102–109. DOI: 10.1016/j.wneu.2021.04.081

[3] Shriver MF, Zeer V, Alentado VJ, Mroz TE, Benzel EC, Steinmetz MP. Lumbar spine surgery positioning complications: A systematic review. Neurosurgical Focus. 2015;**39**(4):E16. DOI: 10.3171/2015.7.FOCUS15268

[4] Berry C, Sandberg DI, Hoh DJ, Krieger MD, McComb JG. Use of Cranial Fixation Pins in pediatric neurosurgery. Neurosurgery. 2008;**62**(4):913–919. DOI: 10.1227/01.neu.0000318177.95288.cb

[5] Agrawal D, Steinbok P. Simple technique of head fixation for image-guided neurosurgery in infants. Child's Nervous System. 2006;**22**(11):1473–1474. DOI: 10.1007/s00381-006-0244-y

[6] Gupta N. A modification of the mayfield horseshoe headrest allowing pin fixation and cranial immobilization in infants and young children. Operative Neurosurgery. 2006;**58**: ONS–E181. DOI: 10.1227/01.NEU.0000193929.11266.87

[7] LoPresti MA, Nguyen J, Lam SK. Pinning in pediatric neurosurgery: The modified rubber stopper technique. Journal of Neurosurgery: Pediatrics. 2020;**26**(1):98–103. DOI: 10.3171/2020.1.PEDS19541

[8] St-Arnaud D, Paquin MJ. Safe positioning for neurosurgical patients. AORN Journal. 2008;**87**(6):1156–1172. DOI: 10.1016/j.aorn.2008.03.004

[9] Magnæs B. Clinical studies of cranial and spinal compliance and the craniospinal flow of cerebrospinal fluid. British Journal of Neurosurgery. 1989;**3**(6):659–668. DOI: 10.3109/02688698908992689

[10] Sattur MG, Patel SJ, Helke KL, Donohoe M, Spiotta AM. Head elevation, cerebral venous system, and intracranial pressure: Review and hypothesis. Stroke: Vascular and Interventional Neurology. 2022;**3**(4). DOI: 10.1161/SVIN.122.000522

[11] Khawari S, Al-Mohammad A, Pandit A, Moncur E, Bancroft MJ, Tariq K, et al. ICP during head movement: Significance of the venous system. Acta Neurochirurgica. 2023;**165**(11):3243–3247. DOI: 10.1007/s00701-023-05620-8

[12] Oberle L, Tatagiba M, Naros G, Machetanz K. Intermittend Pneumatic Venous Thrombembolism (VTE) prophylaxis during neurosurgical procedures. Acta Neurochirurgica. 2024;**166**(1). DOI: 10.1007/s00701-024-06129-4

[13] Qi X, Wang M, Feng K, Ma Y, Zhang D, Guo Y, et al. The effect of intraoperative intermittent pneumatic compression on deep venous thrombosis prophylaxis in patients undergoing elective craniotomy.

Frontiers in Neurology. 2024:15. DOI: 10.3389/fneur.2024.1421977

[14] Wang X, Zhang Y, Fang F, Jia L, You C, Xu P, et al. Comparative efficacy and safety of pharmacological prophylaxis and intermittent pneumatic compression for prevention of venous thromboembolism in adult undergoing neurosurgery: A systematic review and network meta-analysis. Neurosurgical Review. 2020;**44**(2):721–729. DOI: 10.1007/s10143-020-01297-0

[15] Shimizu T, Toyota S, Nakagawa K, Murakami T, Mori K, Kishima H, et al. Retrosigmoid approach in the supine position using ORBEYE: A consecutive series of 14 cases. Neurologia Medico-chirurgica. 2020;**61**(1):55–61. DOI: 10.2176/nmc.tn.2020-0277

[16] Awad AJ, Zaidi HA, Albuquerque FC, Abla AA. Gravity-Dependent Supine Position for the Lateral Supracerebellar Infratentorial Approach. Operative Neurosurgery. 2016;**12**(4):317–325. DOI: 10.1227/NEU.0000000000001327

[17] Balasa A, Hurghis CI, Tamas F, Chinezu R. Patient positioning in neurosurgery, principles and complications. Acta Marisiensis - Seria Medica. 2020;**66**(1):9–14. DOI: 10.2478/amma-2020-0007

[18] Elguindy MM, Bernabei JM, Young JS, Berger MS. Ipsilateral tumor-side-down approach to mesial parietal and occipital gliomas: Patient series. Journal of Neurosurgery: Case Lessons. 2025;**9**(26). DOI: 10.3171/CASE25305

[19] Marotta DA, Brazdzionis J, Fiani B, Duong J, Noel J, Siddiqi J. Perioperative positioning in neurosurgery: A Technical Note on Park Bench Positioning for the obese Patient using the “Arrowhead. Cureus. 2021. DOI: 10.7759/cureus.16932

[20] Spena G, Guerrini F, Grimod G. A modified park bench position: The “Dormeuse” position. Acta Neurochirurgica. 2019;**161**(9):1823–1827. DOI: 10.1007/s00701-019-04013-0

[21] Ausman JI, Malik GM, Dujovny M, Mann R. Three-quarter prone approach to the pineal-tentorial region. Surgical Neurology. 1988;**29**(4):298–306. DOI: 10.1016/0090-3019(88)90161-9

[22] Kwee MM, Ho YH, Rozen WM. The prone position during surgery and its complications: A systematic review and evidence-based guidelines. International Surgery. 2015;**100**(2): 292–303. DOI: 10.9738/INTSURG-D-13-00256.1

[23] Suzuki M, Sobata E, Hatanaka M, Suzuki S, Iwabuchi T, Makiguchi K. Total removal of a Falcotentorial Junction Meningioma by Biparietooccipital craniotomy in the Sea Lion Position: A case report. Neurosurgery. 1984;**15**(5):710–714. DOI: 10.1227/00006123-198411000-00015

[24] Goraksha S, Thakore B, Monteiro J. Sitting position in neurosurgery. Journal of Neuroanaesthesiology and Critical Care. 2019;**07**(02):077–83. DOI: 10.1055/s-0039-1693078

[25] Hurth H, Ebner FH, Clement E, Naros G, Rosenberger P, Kasper EM, et al. The risk of intraoperative venous air embolism from neurosurgical procedures performed in the lounging position: An in-depth analysis of detection, management, and outcomes of 1000 consecutive cases. Journal of

Neurosurgery. 2024:1–11. DOI: 10.3171/2024.5.JNS232449

[26] Fathi AR, Eshtehardi P, Meier B. Patent foramen ovale and neurosurgery in sitting position: A systematic review. British Journal of Anaesthesia. 2009;**102** (5):588–596. DOI: 10.1093/bja/aep063

[27] Korkmaz Dilmen O, Akcil EF, Tureci E, Tunali Y, Bahar M, Tanriverdi T, et al. Neurosurgery in the sitting position: Retrospective analysis of 692 adult and pediatric cases. Turkish Neurosurgery. 2011. DOI: 10.5137/1019-5149.JTN.4974-11.0

[28] Kobayashi S, Sugita K, Tanaka Y, Kyoshima K. Infratentorial approach to the pineal region in the prone position: Concorde position: Technical note. Journal of Neurosurgery. 1983;**58** (1):141–143. DOI: 10.3171/jns.1983.58.1.0141

[29] Takasuna H, TANAKA Y. The modified concorde position with an intraoperative skew head rotation: Technical Note. Neurologia Medico-chirurgica. 2015;**55**(8):680–682. DOI: 10.2176/nmc.tn.2014-0427

[30] Kyoshima K. Arm-down concorde position. Surgical Neurology. 2002;**57** (6):443–445. DOI: 10.1016/s0090-3019(02)00711-5

[31] Datta D, Chatterjee D, Tiwari M. Lingual abscess after Posterior Fossa Surgery: An unusual complication of the concorde position. Asian Journal of Neurosurgery. 2023;**18**(04):810–812. DOI: 10.1055/s-0043-1776299

[32] Ishikawa K, Endo H, Ohtake Y, Watanabe T, Nakamura H. Thoracic spinal cord injury after surgical removal of a ruptured cerebellar arteriovenous malformation in a patient in the Concorde position: Illustrative case. Journal of Neurosurgery: Case Lessons. 2024;**7**(16). DOI: 10.3171/CASE249

[33] Kumar A, Sahu B, Bhattacharjee S, Kulkarni D. Brain death after Concorde positioning for supracerebellar-infratentorial approach: Unanswered questions and lessons learned. Surgical Neurology International. 2012;**3**(1):37. DOI: 10.4103/2152-7806.94036

Chapter 10

Intelligent Perioperative Pathways and Life Expectancy: Neurosurgical Care at the Intersection of Technology, Health Systems and Equity

Ali Msheik, Ruben Peralta, Zeinab Al Mokdad, Nasser Alsaad, Abdulla Illeyan, Muath Hussein, Muhammad Mohsin Khan, Amr El Mohamad and Kazim Mohammed

Abstract

Life expectancy in the twenty-first century increasingly reflects the performance of health systems in delivering complex, high-risk care. As noncommunicable diseases and trauma dominate global mortality, surgical systems, particularly neurosurgery, have become structural determinants of survival and long-term functional independence. This chapter examines the concept of the intelligent perioperative pathway as a systems-level innovation linking predictive analytics, real-time physiological monitoring, and adaptive decision support across preoperative, intraoperative, and postoperative phases of care. We synthesize current evidence on machine learning–based risk stratification, artificial intelligence–augmented intraoperative management, and postoperative early warning systems, and analyze their implications for short-term mortality, quality-adjusted life years, and modeled long-term survival. The chapter further evaluates governance, equity, regulatory, and implementation challenges that shape the translation of technological innovation into population-level longevity gains, particularly in resource-constrained settings. We argue that intelligent perioperative systems reconfigure neurosurgical care from episodic intervention to continuous, data-informed optimization, reducing preventable complications and strengthening functional outcomes. However, sustainable impact depends on ethical oversight, equitable infrastructure development, transparent model validation, and alignment with broader health system strengthening strategies. When responsibly implemented at scale, intelligent perioperative pathways may contribute to measurable improvements in survival, disability reduction, and population life expectancy.

IntechOpen

Keywords: intelligent perioperative care, neurosurgery, artificial intelligence in surgery, health systems engineering, life expectancy, surgical outcomes, digital health equity

1. Introduction

In the twenty-first century, life expectancy depends not only on demographic and preventive advances but increasingly on the capacity of health systems to deliver safe, timely, and complex care. As infectious disease mortality declines, noncommunicable diseases, trauma, and technology-intensive interventions have become dominant determinants of longevity [1]. Surgical systems managing high-risk, time-sensitive conditions now shape survival outcomes. Neurosurgery represents a high-impact domain within modern health systems. Conditions including traumatic brain injury, intracranial tumors, cerebrovascular disease, hydrocephalus, epilepsy, and degenerative spine disorders contribute substantially to premature mortality and long-term disability [2]. Outcomes are highly sensitive to perioperative system performance. Traditionally guided by clinical judgment, protocols, and retrospective audit, perioperative care is now increasingly supported by digital infrastructure, high-frequency monitoring, and advanced analytics. Artificial intelligence (AI) and predictive modeling enable real-time risk stratification and adaptive decision support across the surgical continuum [3]. The intelligent perioperative pathway integrates data-driven tools across preoperative assessment, intraoperative management, and postoperative recovery within a coordinated, learning system. In neurosurgery, such integration reduces preventable complications and improves long-term outcomes [4].

This chapter examines intelligent perioperative pathways in neurosurgery as a systems-level response to contemporary longevity challenges. It analyzes how predictive modeling, real-time monitoring, and adaptive decision support influence perioperative mortality and long-term survival within evolving health systems.

2. Literature search strategy and evidence synthesis approach

This chapter is based on a structured narrative review of the literature examining intelligent perioperative systems, neurosurgical outcomes, predictive analytics, and health systems performance. A targeted literature search was conducted using PubMed/MEDLINE, Scopus, Web of Science, and Google Scholar for publications between January 2000 and January 2026. Search terms included combinations of: "intelligent perioperative care," "artificial intelligence in surgery," "machine learning neurosurgery," "perioperative mortality," "neurocritical care analytics," "early warning systems," "quality-adjusted life years," "surgical systems strengthening," and "digital health equity." Inclusion criteria comprised peer-reviewed original research, systematic reviews, meta-analyses, consensus statements, and high-quality narrative reviews addressing perioperative intelligence, neurosurgical outcomes, AI-enabled risk modeling, or health systems implications. Studies limited to nonclinical simulation environments without clinical relevance were excluded. Priority was given to studies reporting measurable outcomes, including mortality, complication rates, length of stay, quality-adjusted life years (QALYs), disability-adjusted life years (DALYs), or validated predictive performance metrics (area under the curve [AUC], calibration statistics). This chapter synthesizes conceptual, clinical, and systems-level evidence and does not involve original data collection.

3. Life expectancy as a health systems outcome

3.1 Evolution of life expectancy in the twenty-first century

Life expectancy rose dramatically in the twentieth century through advances in sanitation, vaccination, antimicrobial therapy, maternal care, and nutrition. Today, as infectious disease mortality declines, noncommunicable diseases, trauma, neurological disorders, and age-related conditions dominate the global burden of mortality and disability [5]. Life expectancy gains remain uneven across geographic and socioeconomic contexts. High-income countries continue to improve survival largely through advances in cardiovascular care, oncology, critical care, and surgical safety, whereas many low- and middle-income countries (LMICs) face a dual burden of communicable and noncommunicable diseases alongside limited surgical infrastructure and workforce shortages [6]. Recent global disruptions, including pandemics, conflict, and economic instability, underscore the sensitivity of longevity to system resilience and capacity.

Longevity is increasingly assessed not only in years lived but also in years lived with functional independence. Composite metrics such as QALYs and DALYs integrate survival with functional status [7]. In neurosurgical conditions, postoperative cognitive and neurological outcomes directly shape both individual life trajectories and population-level longevity indicators. Life expectancy in the contemporary era increasingly reflects the performance of acute and surgical care systems, alongside preventive health strategies.

3.2 Surgical systems as determinants of population longevity

Surgical conditions represent a major component of the global burden of disease. Trauma, malignancy, obstetric emergencies, congenital anomalies, and neurological disorders often require operative intervention for survival or functional preservation. Accordingly, the quality and accessibility of surgical systems directly influence mortality and long-term disability [8].

Surgical systems are composite infrastructures linking workforce, facilities, supply chains, referral networks, perioperative protocols, and critical care capacity. The perioperative period remains a critical determinant of surgical survival [9]. Thirty-day postoperative mortality constitutes a substantial share of preventable deaths worldwide. Advances in anesthesia safety, infection control, blood management, and complication surveillance have improved outcomes in high-resource settings; integration of predictive analytics and real-time monitoring represents a further systems-level refinement. Surgical systems function not merely as service delivery platforms but as structural determinants of population longevity [10].

3.3 Neurosurgical conditions and premature mortality

Neurological and neurosurgical disorders constitute a major source of premature mortality and long-term disability. Traumatic brain injury, intracranial hemorrhage, stroke, brain and spinal tumors, hydrocephalus, epilepsy, and spinal cord compression frequently require urgent intervention to prevent irreversible secondary injury. Delayed or suboptimal management in these conditions directly translates into avoidable mortality and sustained neurological impairment [11].

The impact on life expectancy extends beyond perioperative survival. Survivors of severe brain injury or hemorrhage often face persistent cognitive deficits, epilepsy, psychiatric complications, and elevated cardiovascular risk, each contributing to excess long-term mortality. In resource-constrained settings, delayed diagnosis, limited imaging access, and inadequate monitoring further compound this burden [12].

Neurosurgical outcomes are acutely sensitive to time-to-intervention, intraoperative precision, and postoperative critical care. Deviations in intracranial pressure, oxygenation, or hemodynamic stability can cause irreversible injury. By reducing modifiable perioperative risks and strengthening neurosurgical systems, health systems can influence both immediate survival and long-term functional independence. Neurosurgical care bridges acute intervention and population health [13]. As illustrated in **Figure 1**, perioperative optimization influences downstream functional and survival trajectories.

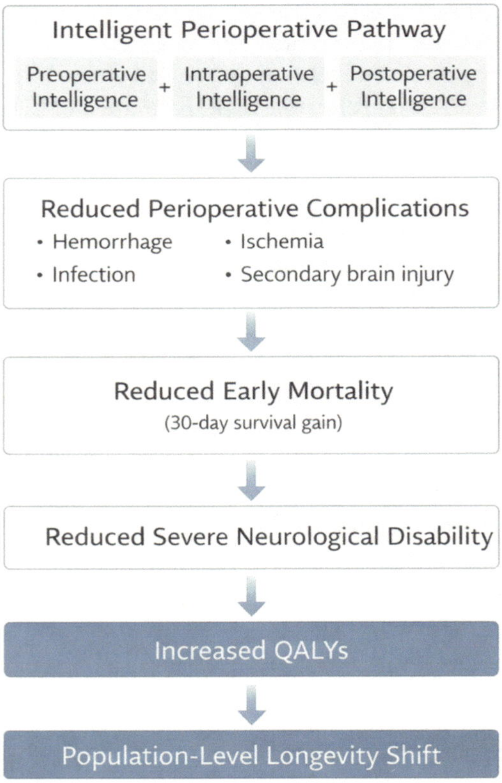

Figure 1.
Conceptual cascade linking intelligent perioperative care to population longevity.

4. The concept of the intelligent perioperative pathway

The intelligent perioperative pathway reconfigures surgical care as a continuous, data-informed system that integrates digital infrastructure with clinical expertise. Instead of treating preoperative, intraoperative, and postoperative phases as discrete events, it frames the surgical journey as a connected continuum [14].

4.1 Defining intelligence in surgical systems

Intelligence in surgical systems denotes the capacity to collect, interpret, and act upon clinical data to improve decision quality and outcomes. It encompasses predictive modeling, real-time awareness, adaptive response, and iterative refinement. Predictive analytics estimate perioperative risk trajectories from demographic, clinical, imaging, and physiological variables; integrated monitoring detects early deterioration; decision-support mechanisms translate analysis into actionable guidance; and outcome feedback informs iterative model refinement. Intelligence augments rather than replaces clinical judgment [15]. In neurosurgery, this may involve anticipating intracranial pressure instability, optimizing cerebral perfusion, or identifying patients at elevated postoperative risk. Effective systems, therefore, require transparency, validation, and accountability to ensure safe and equitable use.

4.2 Digital infrastructure and data ecosystems

Intelligent perioperative systems depend on interoperable digital infrastructure. Electronic health records (EHRs), imaging repositories, laboratory systems, anesthesia platforms, and high-frequency monitors generate structured and unstructured data that must be harmonized through standardized terminology, interoperable architecture, and secure real-time analytics [16].

In neurosurgery, relevant inputs include imaging, comorbidity profiles, neurophysiological signals, hemodynamic data, and postoperative assessments. Integrating these heterogeneous streams enables dynamic risk modeling across care phases. Disparities in digital capacity, particularly in low- and middle-income settings with fragmented documentation and limited connectivity, pose significant implementation challenges. Equitable deployment, therefore, requires infrastructure investment and context-sensitive adaptation [17].

4.3 Integration across the perioperative continuum

The defining feature of the intelligent pathway is integration across preoperative, intraoperative, and postoperative phases. Predictive modeling informs surgical planning, real-time monitoring supports intraoperative stability, and early warning systems (EWS) detect postoperative deterioration before overt decompensation [18]. Continuous information flow enables anticipatory rather than reactive care, with outcome feedback supporting iterative refinement. When effectively governed, such integration strengthens perioperative system performance. The intelligent perioperative pathway functions as a coordinated system for perioperative optimization [19].

5. Preoperative intelligence: Risk stratification and predictive modeling

Preoperative decision-making is a critical phase in the neurosurgical pathway, encompassing operative indication, individualized risk assessment, and alignment of treatment with expected survival and functional outcomes. Although guided by clinical expertise, scoring systems, imaging, and multidisciplinary review, traditional assessment struggles to capture complex, nonlinear risk interactions [20]. Preoperative intelligence restructures this process through predictive modeling and integrated data analysis. Accurate preoperative risk stratification improves survival by enabling targeted optimization, realistic prognostic communication, and appropriate critical care allocation. Even modest reductions in perioperative complications can shift survival trajectories in conditions such as intracranial tumors, vascular malformations, and traumatic brain injury [21].

5.1 Machine learning in neurosurgical risk assessment

Machine learning (ML) techniques enable the modeling of complex perioperative risk profiles. Unlike traditional regression, they process high-dimensional data and detect interactions among demographics, comorbidities, imaging features, laboratory values, and physiological variables. In neurosurgery, such models have been applied to predict hemorrhage, infection, prolonged ventilation, readmission, neurological decline, and short-term mortality [22].

Quantified tumor morphology, edema patterns, vascular proximity, and midline shift, alongside frailty indices and inflammatory markers, enhance the discrimination of adverse outcome risk. Clinical utility depends on transparency, interpretability, and workflow integration. Opaque predictions undermine clinician trust; explainable models and clear uncertainty communication are, therefore, essential [23]. When rigorously validated and calibrated, ML-based risk stratification supports anticipatory planning and reduces preventable complications. The predictive modeling workflow is illustrated in **Figure 2**.

5.2 Personalized surgical planning

Beyond risk estimation, preoperative intelligence informs operative timing, surgical approach, anesthesia strategy, monitoring requirements, and postoperative care levels. In high-risk patients, predictive analytics guide prehabilitation, comorbidity optimization, and early neurocritical care involvement [24].

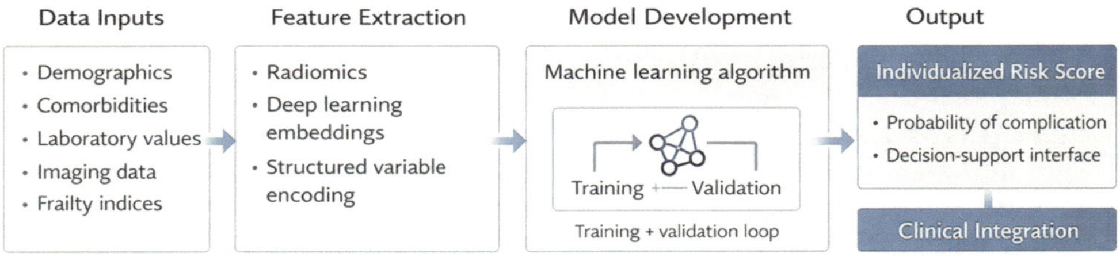

Figure 2.
Predictive modeling architecture in preoperative intelligence.

Personalization also strengthens shared decision-making. Transparent risk projections allow patients to weigh survival benefits against procedural risks and neurological sequelae. In malignant tumors or complex vascular lesions, individualized survival modeling informs choices between aggressive intervention and palliation, aligning technical planning with patient values. Individualized planning also improves resource allocation [25]. Anticipating intensive care needs, blood product use, and length of stay enhances operational efficiency and reduces strain on constrained systems. Scaled across populations, such efficiencies contribute to broader longevity gains.

5.3 Bias, calibration, and validation challenges

Predictive modeling remains vulnerable to bias, miscalibration, and limited generalizability. Algorithms trained in high-resource settings may perform poorly in different demographic or resource contexts, potentially amplifying existing inequities. External validation across diverse populations is, therefore, essential before deployment. Calibration, the alignment between predicted and observed outcomes, requires continuous monitoring. As practices evolve, model performance may degrade (model drift), necessitating auditing, recalibration, and transparent reporting. Governance frameworks must clarify accountability for algorithm-assisted decisions. Predictive tools should support, not supplant, clinical judgment, particularly in high-stakes neurosurgical contexts [26]. Common sources of bias and corresponding mitigation strategies are summarized in **Table 1**.

6. Intraoperative intelligence: Real-time monitoring and adaptive decision support

The intraoperative phase demands continuous physiological stability. Intraoperative intelligence operates synchronously with clinical events, integrating physiological signals and procedural variables to identify risk patterns before overt complications. The interaction between monitoring systems and clinician decision-making is depicted in **Figure 3**.

6.1 Advanced physiological surveillance

Modern neurosurgery generates high-resolution data from multimodal monitoring, including invasive hemodynamics, capnography, cerebral

Bias type	Source	Impact	Mitigation strategy
Sampling bias	High-resource training data	Poor generalizability	External validation
Measurement bias	Incomplete datasets	Miscalibration	Data harmonization
Structural bias	Historical inequities	Outcome distortion	Subgroup analysis
Model drift	Changing clinical practice	Performance degradation	Continuous auditing

Table 1.
Bias in predictive modeling and mitigation strategies.

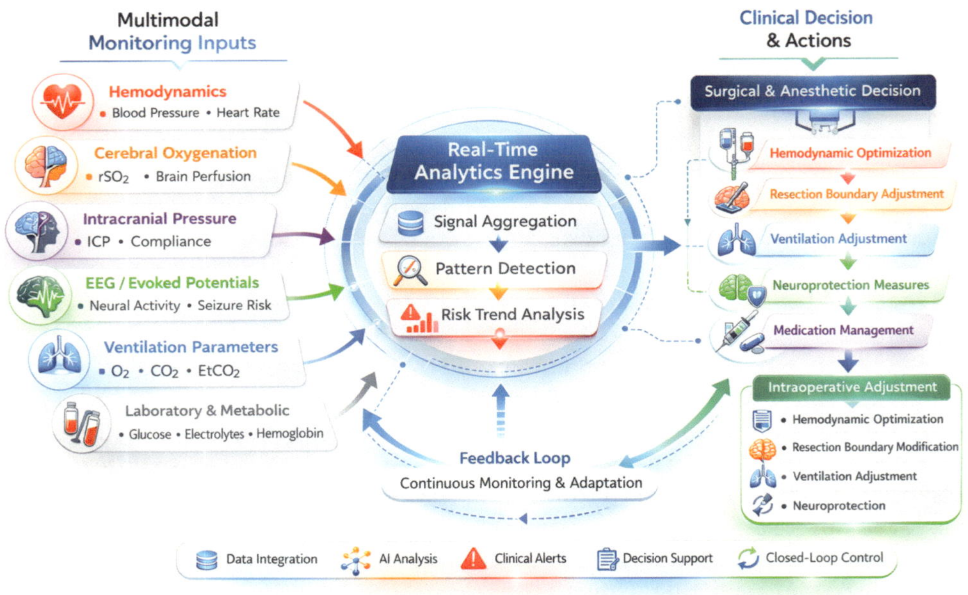

Figure 3.
Real-time monitoring and human–machine collaboration.

oximetry, intracranial pressure, electroencephalography, evoked potentials, and anesthetic depth indices. Together, these parameters provide a multidimensional representation of cerebral and systemic physiology [27]. Advanced surveillance aggregates and analyzes these signals to detect patterns preceding deterioration. Gradual declines in cerebral oxygenation with hemodynamic variability may signal ischemia, while neurophysiological changes during tumor resection or spinal decompression can indicate evolving neural compromise. The limitation lies in human cognitive bandwidth rather than data availability. High-frequency monitoring can exceed attentional capacity during complex procedures. Intelligent analytics assist by filtering noise, prioritizing trends, and reducing alarm fatigue while enhancing early detection of meaningful deviations [28].

6.2 AI-augmented intraoperative decision-making

Intraoperative AI applies real-time pattern recognition to support adaptive surgical and anesthetic decision-making. **Figure 3** details this section.

6.3 Human–machine collaboration in the operating room

Effective intraoperative intelligence depends on seamless collaboration between clinicians and digital systems. In the operating room, a complex sociotechnical environment, technology must support rather than disrupt workflow. Collaboration requires transparent algorithms, intuitive interfaces, and defined accountability [29]. Clinicians must understand the basis and uncertainty of alerts; explainable models and clear visualization enhance interpretability and trust. Training and simulation are essential to prevent overreliance or inappropriate dismissal of alerts. Systems require rigorous testing in simulated and real-world settings to assess usability and unintended consequences. Human expertise remains indispensable for contextual interpretation, ethical reasoning, and ultimate accountability. When appropriately integrated, human–machine collaboration reduces preventable harm and strengthens neurological preservation, with downstream implications for longevity [30].

7. Postoperative intelligence: EWSs and long-term outcome optimization

The postoperative phase determines whether operative success translates into a durable recovery or secondary deterioration. Postoperative instability directly influences survival and functional recovery.

7.1 Complication prediction and early detection

Early warning systems form the foundation of postoperative intelligence. By integrating physiological parameters – heart rate, blood pressure, respiratory rate, oxygen saturation, temperature, and neurological status – into composite risk indices, these systems detect early signals of deterioration. In neurosurgical populations, additional inputs such as pupillary reactivity, Glasgow Coma Scale trends, intracranial pressure, and imaging findings enhance predictive accuracy [31]. Machine learning–enhanced models extend beyond threshold-based scoring by identifying nonlinear patterns that precede overt clinical decline. Subtle shifts in cerebral perfusion pressure, laboratory trends, or neurological examination may signal impending deterioration before classical signs emerge. Predictive systems can also identify risks of infection, thromboembolism, or unplanned intensive care readmission. The clinical value lies in shortening the interval between physiological deviation and therapeutic response. Earlier intervention reduces preventable mortality and neurological disability. Accuracy, calibration, and minimization of false alarms remain essential to prevent alert fatigue and preserve clinician trust [32].

7.2 Critical care optimization

Neurosurgical patients frequently require specialized neurocritical care to prevent secondary brain injury. Maintenance of intracranial pressure, cerebral perfusion, ventilation parameters, sedation balance, and metabolic stability is central to outcome preservation. Intelligent systems support this process through continuous multimodal data integration and adaptive decision support [33]. Predictive analytics can estimate individualized blood pressure targets based on autoregulatory status, recommend ventilation adjustments to maintain appropriate carbon dioxide levels, and detect evolving organ dysfunction through trend analysis. Integration of intracranial pressure monitoring, brain tissue oxygenation, electroencephalography, and systemic hemodynamics enables more precise physiologic control. Beyond physiological management, predictive tools improve operational efficiency by estimating length of stay, guiding step-down timing, and optimizing resource allocation. Improved neurocritical care performance reduces postoperative mortality and supports system resilience [34].

7.3 Functional recovery and longitudinal data integration

Postoperative intelligence extends beyond immediate survival to long-term functional recovery. Survivors of neurosurgical intervention may experience cognitive impairment, motor deficits, seizures, psychiatric sequelae, or chronic pain, each influencing quality-adjusted life expectancy [35]. Longitudinal data integration, through structured follow-up, telemedicine platforms, wearable sensors, and patient-reported outcomes, enables tracking of recovery trajectories. Machine learning models can identify patients at risk of delayed deterioration, rehospitalization, or functional decline, prompting earlier rehabilitation, medication adjustment, or psychosocial intervention. Incorporating long-term outcome data into perioperative databases establishes feedback loops that refine predictive models across all surgical phases [36].

8. Impact on mortality, functional outcomes, and life expectancy

The value of intelligent perioperative pathways lies in demonstrable improvements in mortality and functional outcomes. The relationship between the perioperative phase, intermediate outcomes, and long-term health impact is summarized in **Table 2**.

8.1 Short-term mortality reduction

Thirty-day postoperative mortality remains a central indicator of surgical quality and system performance. In neurosurgery, early mortality frequently results from modifiable factors, including uncontrolled intracranial hypertension, hemorrhage, ischemia, infection, cardiopulmonary instability, and delayed recognition of deterioration. Preoperative risk stratification enables targeted optimization and appropriate allocation of intensive care resources [37]. Intraoperative analytics support hemodynamic stability and early detection of neural compromise. Postoperative surveillance systems shorten the interval between physiological deviation and therapeutic intervention. In time-sensitive conditions, such as traumatic brain injury and intracranial hemorrhage, earlier stabilization and anticipatory management shift early survival curves and preserve the potential for long-term recovery [38].

Large database analyses of neurosurgical populations report 30-day mortality rates ranging from 2% to 8%, depending on pathology and system capacity. Predictive modeling and enhanced postoperative surveillance have been associated, in observational studies, with relative mortality reductions of approximately 10–20%

Intervention phase	Immediate impact	Intermediate outcome	Long-term impact
Preoperative	Risk stratification	Reduced complications	Survival gain
Intraoperative	Physiological stabilization	Reduced deficit	Functional preservation
Postoperative	Early detection	Reduced disability	QALY increase

Table 2.
Perioperative intelligence and health outcome domains.

in high-risk surgical cohorts. Even a 1–2% absolute reduction in early postoperative mortality at scale represents a substantial population-level survival gain in high-volume neurosurgical systems.

8.2 QALYs and DALYs

Life expectancy alone incompletely captures the impact of neurosurgical care. Survivors of severe neurological injury may experience persistent cognitive, motor, or psychological impairments affecting independence and societal participation. Composite metrics such as QALYs and DALYs integrate survival duration with functional status, providing a more comprehensive assessment of health impact. Gains arise from both extended survival and preserved functional capacity [39]. Health economic modeling studies in surgical oncology and critical care suggest that the prevention of major neurological deficits may yield QALY gains ranging from 0.5 to 3.0 per patient, depending on baseline functional status and life expectancy. In neurosurgical populations, the avoidance of severe disability through early detection of complications or optimized hemodynamic control may therefore produce measurable improvements in quality-adjusted survival beyond crude mortality reduction.

8.3 Modeling long-term survival gains

Long-term survival gains require rigorous modeling. Survival analyses, including Kaplan–Meier estimation and Cox proportional hazards modeling, can compare cohorts managed with conventional versus intelligence-augmented pathways. Simulation models incorporating complication rates, functional outcomes, and rehabilitation trajectories allow estimation of population-level longevity shifts [40]. Long-term survival improvements often arise indirectly through the prevention of secondary complications. Avoidance of severe neurological deficits reduces chronic immobility, recurrent hospitalization, cardiovascular deconditioning, and psychosocial decline, each associated with excess mortality. Similarly, optimized early management of intracranial pathology prevents progressive deterioration that would otherwise shorten lifespan. Modeling must account for heterogeneity in disease severity, comorbidities, demographic context, and health system capacity. External validation across diverse settings remains essential to ensure generalizability and avoid overestimation of benefits [41].

9. Ethical governance and responsible innovation

Impact depends on equitable implementation, governance, data stewardship, and sustained human oversight. Without inclusive policy design, disparities in workforce distribution, digital infrastructure, and critical care capacity, particularly in low- and middle-income settings, may widen existing inequities [42]. The integration of AI and predictive analytics into neurosurgical care introduces governance challenges proportional to its clinical impact. The multilevel governance structure required for safe implementation is shown in **Figure 4**.

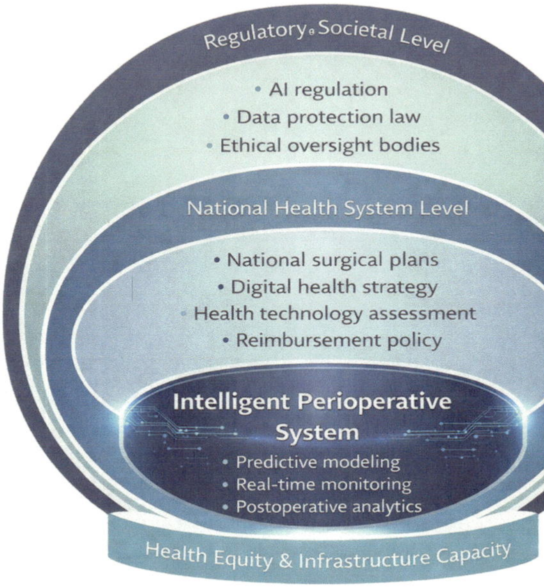

Figure 4.
Governance and system integration framework for intelligent perioperative care.

9.1 Transparency, explainability, and accountability

Transparency is foundational to the responsible use of AI in perioperative care. Clinicians must understand the origin, structure, and limitations of predictive models that inform decision-making [3]. Opaque "black-box" outputs compromise professional autonomy and complicate risk communication with patients. Explainable modeling approaches and clear visualization of risk drivers enhance interpretability and support responsible integration into clinical workflows. Accountability structures must delineate responsibility among developers, institutions, and clinicians. Governance mechanisms should specify requirements for model validation, recalibration, performance auditing, and incident review. Continuous evaluation of discrimination, calibration, and subgroup equity is essential to ensure safety over time. Where predictive analytics materially influence surgical planning or risk communication, patients should be informed of their use. Transparency in this context strengthens shared decision-making and reinforces respect for autonomy [43].

9.2 Data stewardship and privacy

Intelligent perioperative systems depend on extensive clinical datasets, including imaging, physiological monitoring, and longitudinal health records. Responsible data stewardship requires robust privacy safeguards, cybersecurity protections, and clear governance policies governing access, storage, and secondary use. Data minimization, encryption, controlled access, and rigorous de-identification are necessary but not sufficient safeguards, particularly in high-dimensional datasets where re-identification risk persists. Institutions must adopt layered security strategies and ensure compliance with jurisdictional regulations governing data sovereignty and cross-border transfer. Equitable data governance is especially important in international collaborations. Populations contributing data must benefit from resulting innovations, and partnerships should avoid extractive models that reinforce structural inequities [44].

9.3 Human oversight and professional responsibility

Despite advances in computational modeling, human oversight remains central to perioperative decision-making. Algorithms identify patterns and generate risk estimates, but they cannot fully account for contextual nuance, patient values, or ethical complexity. Clinicians retain ultimate responsibility for interpretation and action [45]. Effective oversight includes monitoring for model drift, critically evaluating algorithmic recommendations, and maintaining the capacity to override automated outputs when clinical judgment warrants deviation. Education in digital literacy and algorithmic limitations is essential to prevent overreliance or inappropriate dismissal. Institutional governance structures, such as multidisciplinary oversight committees, support responsible deployment by integrating clinical, technical, and ethical expertise. Through sustained oversight, intelligent systems can enhance rather than displace professional responsibility [46].

10. Equity and implementation in diverse health system contexts

Access to safe neurosurgical care and digital infrastructure remains uneven across health systems. Implementation must, therefore, align technological innovation with broader health system strengthening rather than treating intelligent tools as isolated upgrades.

10.1 Digital divides and infrastructure gaps

Substantial disparities in digital capacity constrain equitable deployment. High-resource institutions often operate integrated EHRs, advanced monitoring platforms, secure data storage, and real-time analytics. In contrast, many facilities, particularly in rural and resource-limited regions, rely on fragmented documentation systems, limited connectivity, and constrained technical support [47]. These gaps affect both implementation feasibility and data quality. Incomplete or inconsistent data reduce model reliability and increase the risk of algorithmic bias. Systems designed for high-bandwidth environments may prove unsustainable where infrastructure is unstable or maintenance capacity is limited. Addressing digital divides requires scalable, interoperable, and context-adapted solutions. Investment in foundational digital infrastructure, secure data systems, standardized documentation, and reliable connectivity forms a prerequisite for equitable perioperative intelligence [48].

10.2 LMIC considerations

Low- and middle-income countries bear a disproportionate burden of untreated neurosurgical diseases, including traumatic brain injury, hydrocephalus, and intracranial infections. Workforce shortages, delayed referral systems, and limited neurocritical care capacity contribute to elevated perioperative mortality. In these settings, intelligent perioperative tools may support risk stratification, teleconsultation, and EWSs that partially mitigate workforce constraints. However, algorithms trained exclusively in high-income contexts may lack validity where epidemiology, comorbidity profiles, and resource availability differ. Local validation and contextual adaptation are therefore essential [49]. Cost-effectiveness is particularly salient in resource-constrained systems. Intelligent tools must demonstrate measurable reductions in complications, length of stay, and disability relative to competing health priorities. When adapted appropriately, such systems can strengthen surgical capacity and contribute to improved survival trajectories without exacerbating inequity [49].

10.3 Sustainable capacity building

Sustainable implementation requires long-term capacity development rather than one-time technological deployment. Capacity building encompasses workforce training, digital literacy, biomedical engineering support, regulatory development, and institutional governance. Collaborative partnership models emphasizing co-development, local data stewardship, and reciprocal benefit promote equitable innovation. Investment in maintenance infrastructure, cybersecurity expertise, and continuous professional education ensures durability beyond initial adoption [50]. Embedding intelligent perioperative systems within resilient health systems amplifies their impact. When aligned with equitable access and quality improvement strategies, digital augmentation of neurosurgical care can reduce preventable mortality and disability across diverse contexts. Aggregated at scale, such improvements support more equitable gains in life expectancy. Addressing structural inequities is inseparable from regulatory and policy frameworks that govern intelligent system deployment [51].

11. Policy, regulation, and system-level integration

Deployment must meet standards for safety, transparency, and accountability comparable to other medical technologies. Sustainable adoption depends not only on technical validation but also on integration within national digital health strategies, surgical system planning, and institutional governance structures. Integration within national surgical and digital health strategies prevents fragmentation of care delivery [52]. As shown in **Figure 4**, sustainable deployment requires alignment across institutional, national, and regulatory layers. Key system domains, policy instruments, and indicators of readiness are outlined in **Table 3**.

11.1 Regulatory frameworks for AI in surgery

AI-enabled systems challenge conventional regulatory paradigms, particularly when models evolve through iterative updating. Regulatory frameworks must, therefore, accommodate adaptive software while preserving rigorous predeployment validation and postdeployment surveillance. Regulatory oversight must include transparent validation, postdeployment monitoring, and subgroup performance auditing [53].

Domain	Key requirements	Policy instruments	Indicators of readiness
Digital infrastructure	Interoperable EHR, data storage, connectivity	National digital health strategy	Percentage of facilities with interoperable systems
Clinical governance	Validation protocols, model auditing	Professional guidelines	Model recalibration frequency
Regulation	AI approval pathways, data protection	Regulatory frameworks	Compliance certification
Workforce	Digital literacy, training programs	Education policy, CME mandates	Trained personnel ratio
Equity	LMIC adaptation, access safeguards	Equity-focused funding	Regional outcome parity

Table 3.
Policy levers and system requirements for intelligent perioperative implementation.

11.2 Health economics and cost-effectiveness

Long-term sustainability depends on demonstrable economic value. Investments in digital infrastructure, analytics platforms, cybersecurity, and workforce training must yield measurable clinical benefits relative to cost. Cost-effectiveness analyses should incorporate reductions in complication rates, shortened hospital stays, decreased readmissions, and improved functional outcomes. In neurosurgical populations, prevention of severe disability generates substantial downstream savings through reduced rehabilitation costs, long-term care dependency, and productivity loss [54]. In resource-constrained settings, opportunity costs require careful evaluation. Scalable, interoperable solutions that deliver measurable improvements in survival and QALYs offer the strongest justification for adoption. Economic evaluations in perioperative and critical care settings suggest that interventions reducing major complications by even 5–10% may achieve favorable cost-effectiveness ratios, particularly when associated with QALY gains exceeding 0.5 per treated high-risk patient. In neurosurgical populations, prevention of severe long-term disability generates substantial downstream societal and healthcare cost savings, strengthening the economic rationale for responsible implementation.

11.3 Public trust and societal implications

Sustainable adoption depends on public trust. Patients must have confidence that AI-enabled perioperative care is safe, equitable, and ethically governed. Transparent communication regarding data use, algorithmic support, and institutional safeguards strengthens legitimacy. The societal implications extend to professional roles and training paradigms. Surgeons and anesthesiologists increasingly operate within data-rich environments requiring digital literacy and interdisciplinary collaboration [55]. Educational frameworks must evolve accordingly. Equitable distribution of technological benefits remains central. If intelligent systems disproportionately advantage well-resourced institutions, public confidence may erode. Policy frameworks must, therefore, prioritize inclusive deployment to ensure that longevity gains are broadly shared.

12. Future directions in intelligent neurosurgical systems

Advances in computational modeling, wearable biosensors, teleneurosurgery, and remote monitoring may extend perioperative intelligence beyond hospital boundaries into longitudinal care environments. Multimodal data fusion, including imaging, physiological monitoring, genomics, and patient-reported outcomes, will enable increasingly individualized surgical pathways. Federated learning approaches may enhance generalizability while preserving data privacy by allowing distributed model training without centralized data pooling. Emerging concepts, such as patient-specific computational modeling and digital simulation environments, may further refine operative planning and risk prediction. Prospective multicenter evaluation remains essential to determine whether these innovations translate into sustained reductions in mortality and disability. Intelligent neurosurgical systems should evolve as continuously validated learning ecosystems embedded within accountable health systems.

This chapter is a structured narrative synthesis and does not present original prospective or retrospective data. Conclusions regarding mortality reduction, QALY gains, and long-term survival impact are based on extrapolation from published observational studies, modeling analyses, and related surgical literature rather than direct interventional trials of fully integrated intelligent perioperative pathways. Heterogeneity in neurosurgical pathology, resource availability, and digital infrastructure limits the generalizability of reported effect sizes. Furthermore, many cited studies evaluate discrete components (predictive modeling or EWS) rather than fully integrated perioperative ecosystems. Prospective multicenter trials and long-term population-level evaluations are required to quantify the true magnitude of survival and functional gains attributable to intelligent perioperative systems.

13. Conclusions

Intelligent perioperative pathways may reconfigure neurosurgical care through integrated predictive modeling, real-time monitoring, and adaptive decision support. By strengthening risk stratification, intraoperative stability, and postoperative surveillance, they address modifiable determinants of surgical outcomes. When embedded within resilient and ethically governed health systems, such pathways improve quality-adjusted survival and contribute to population longevity. Their value will be measured by sustained reductions in preventable mortality and disability across populations.

Abbreviations

AI	Artificial intelligence
DALYs	Disability-adjusted life years
EHR	Electronic health record
EWS	Early warning system
LMICs	Low- and middle-income countries
ML	Machine learning
QALYs	Quality-adjusted life years

Author contributions

A.M. conceived the chapter concept and overall structure. A.M., R.P., and Z. A.M. contributed to the development of the conceptual framework and systems-level analysis. N.A., A.I., M.H., M.M.K., A.E.M., and K.M. contributed to clinical content development, technical review, and domain-specific refinement in neurosurgical practice and perioperative systems. Z.A.M. provided expertise on ethical governance and responsible innovation. All authors contributed to drafting, critical revision, and approval of the final chapter. All authors agree to be accountable for the content of this work.

Funding

The authors received no specific funding for the preparation of this chapter. No external financial support influenced the content, analysis, or conclusions presented herein.

Ethical statement

This chapter is a conceptual and analytical work and does not involve human participants, patient data, or animal research. Ethical approval was, therefore, not required.

Data availability statement

No primary datasets were generated or analyzed for this chapter. All conceptual frameworks and analyses are based on publicly available literature and established methodological principles.

Acknowledgments

The authors acknowledge the contributions of colleagues in neurosurgery, critical care, and health systems research, whose interdisciplinary collaboration informed the development of the concepts presented in this work.

Conflict of Interest

The authors declare no conflict of interest related to the subject matter of this chapter.

Disclosure of AI-enabled tools

Portions of this chapter were developed with the assistance of AI-enabled language refinement tools to improve clarity, structure, and editorial precision. All conceptual content, analytical framing, and scholarly interpretation were developed and critically reviewed by the authors. The authors assume full responsibility for the integrity and originality of the work.

Author details

Ali Msheik[1*], Ruben Peralta[2], Zeinab Al Mokdad[3], Nasser Alsaad[1], Abdulla Illeyan[1], Muath Hussein[1], Muhammad Mohsin Khan[1], Amr El Mohamad[1] and Kazim Mohammed[1]

1 Neuroscience Institute, Hamad Medical Corporation, Doha, Qatar

2 Hamad Medical Corporation, Doha, Qatar

3 Lebanese University, Faculty of Medical Sciences, Beirut, Lebanon

*Address all correspondence to: amsheik@hamad.qa

References

[1] Woo J, Marmot M. Non-Communicable diseases, longevity, and health span: A Hong Kong perspective. International Journal of Environmental Research and Public Health. 2025;**22**(3):359. DOI: 10.3390/ijerph22030359

[2] Roy S, Awuah WA, Ahluwalia A, Adebusoye FT, Ferreira T, Tan JK, Bharadwaj HR, Tenkorang PO, Abdul-Rahman T, Papadakis M. Current trends and challenges: The landscape of perioperative mortality in intracranial surgeries in low- and middle-income settings: A narrative review. Health Science Reports. 2024;**7**(1):e1838. DOI: 10.1002/hsr2.1838

[3] Han L, Char DS, Aghaeepour N; Stanford Anesthesia AI Working Group. Artificial Intelligence in Perioperative Care: Opportunities and Challenges. Anesthesiology. 2024;**141**(2):379–387. DOI: 10.1097/ALN.0000000000005013

[4] Sugiyama T, Sugimori H, Tang M, Fujimura M. Artificial intelligence for patient safety and surgical education in neurosurgery. JMA J. 2025;**8**(1):76–85. DOI: 10.31662/jmaj.2024-0141

[5] GBD, Mortality and Causes of Death Collaborators. Global, regional, and national life expectancy, all-cause mortality, and cause-specific mortality for 249 causes of death, 1980-2015: A systematic analysis for the Global Burden of Disease Study 2015. Lancet. 2016;**388**(10053)1459–1544: DOI: 10.1016/S0140-6736(16)31012-1. Erratum in: L Fernandes Prabhu D, Gurupur V, Stone A, Trader E. Integrating Artificial Intelligence, Electronic Health Records, and Wearables for Predictive, Patient-Centered Decision Support in Healthcare. Healthcare (Basel). 2025 Oct 30;**13**(21):2753. DOI: 10.3390/healthcare13212753. PMID: 41228120; PMCID: PMC12607345.ancet. 2017 Jan 7;389(10064)e1. DOI: 10.1016/S0140-6736(16)32605-8

[6] Asefa ET, Gemmechu TD, A/Fogi MM, Tukeni KN. Increase in cardiovascular disease mortality in low- and middle-income countries: A time for action. Ethiopian Journal of Health Sciences. 2025;**35**(1):1–2. DOI: 10.4314/ejhs.v35i1.1

[7] Nittala MR, Kanakamedala MR, Mundra E, Woods 3rd WC, Smith ML, Hamilton RD, Jefferson GD, Jackson L, Packianathan S, Vijayakumar S. Quality-Adjusted life years and disability-adjusted life years are better with concurrent chemoradiation therapy than induction chemotherapy followed by chemoradiation therapy in nasopharyngeal carcinoma. Cureus. 2021;**13**(1):e13022. DOI: 10.7759/cureus.13022

[8] Maswime S, Jayaraman S, Alaba O, Robalo M. Universal access to surgical care-A global public health priority. PLOS Glob Public Health. 2025;**5**(4):e0004326. DOI: 10.1371/journal.pgph.0004326

[9] Glasbey JC, Ademuyiwa AO, Chu K, Dare A, Harrison E, Hutchinson P, Hyman G, Lawani I, Martin J, Martinez L, Meara J, Reddy KS, Sullivan R. Building resilient surgical systems that can withstand external shocks. BMJ Glob Health. 2024;**9**(Suppl 4):e015280. DOI: 10.1136/bmjgh-2024-015280

[10] Harfaoui W, Alilou M, El Adib AR, Zidouh S, Zentar A, Lekehal B, Belyamani L, Obtel M. Patient safety in anesthesiology: Progress, challenges, and prospects. Cureus. 2024;**16**(9): e69540. DOI: 10.7759/cureus.69540

[11] Kołodziejczyk B, Damps M, Żmudka K, Mandera M. Long-term neurological outcomes of severe traumatic brain injury in the intensive care unit. Frontiers in Pediatrics. 2025;**13**:1582551. DOI: 10.3389/fped.2025.1582551

[12] A DET, Bruni S, Bonavita J, Zadra A, Ciavarella M, Cannavò G, Saviola D. Long-term life expectancy in severe traumatic brain injury: A systematic review. European Journal of Physical and Rehabilitation Medicine. 2024;**60**(5):810–821. DOI: 10.23736/S1973-9087.24.08461-2

[13] Sultan A, Khan MF, Sohaib M, Shamim F. Clinical characteristics and outcomes of neurosurgical patients at a level III intensive care unit in Pakistan: A Retrospective Cohort Study. Cureus. 2024;**16**(1):e52990. DOI: 10.7759/cureus.52990

[14] Feng Z, Bhat RR, Yuan X, Freeman D, Baslanti T, Bihorac A, Li X. Intelligent perioperative system: Towards real-time big data analytics in surgery risk assessment. DASC-PICom-DataCom-CyberSciTech 2017 : 2017 IEEE 15th International Conference on Dependable, Autonomic and Secure Computing ; 2017 IEEE 15th International Conference on Pervasive Intelligence and Computing ; 2017 IEEE 3rd International. 2017;**2017**:1254–1259. DOI: 10.1109/DASC-PICom-DataCom-CyberSciTec.2017.201

[15] Colborn K, Brat G, Callcut R. Predictive analytics and artificial intelligence in surgery-opportunities and risks. JAMA Surgery. 2023;**158**(4):337–338. DOI: 10.1001/jamasurg.2022.5444

[16] Fernandes Prabhu D, Gurupur V, Stone A, Trader E. Integrating artificial intelligence, electronic health records, and wearables for predictive, patient-centered decision support in healthcare. Healthcare (Basel). 2025;**13**(21):2753. DOI: 10.3390/healthcare13212753

[17] Szmyd B, Podstawka M, Wiśniewski K, Zaczkowski K, Puzio T, Tomczyk A, Wojciechowski A, Jaskólski DJ, Bobeff EJ. AI-driven innovations in neuroradiology and neurosurgery: Scoping review of current evidence and future directions. Cancers (Basel). 2025;**17**(16):2625. DOI: 10.3390/cancers17162625

[18] Magouliotis DE, Sicouri N, Ramlawi L, Baudo M, Androutsopoulou V, Sicouri S. Artificial intelligence in adult cardiovascular medicine and surgery: Real-World deployments and outcomes. Journal of Personalized Medicine. 2026;**16**(2):69. DOI: 10.3390/jpm16020069

[19] Kim J, Lui B, Goldstein PA, Rubin JE, White RS, Jotwani R. From data to decisions: Harnessing multi-agent systems for safer, smarter, and more personalized perioperative care. Journal of Personalized Medicine. 2025;**15**(11):540. DOI: 10.3390/jpm15110540

[20] Vintimilla Rivadeneira V, Leon-Rojas JE. Advances in preoperative and intraoperative technologies for safe resection of gliomas in cognitive regions. Cancers (Basel). 2025;**17**(24):3890. DOI: 10.3390/cancers17243890

[21] Yoon HK, Yang HL, Jung CW, Lee HC. Artificial intelligence in perioperative medicine: A narrative review. Korean Journal of Anesthesiology. 2022;**75**(3):202–215. DOI: 10.4097/kja.22157

[22] Hassan AM, Rajesh A, Asaad M, Nelson JA, Coert JH, Mehrara BJ, Butler CE. Artificial Intelligence and machine learning in prediction of surgical complications: Current state, applications, and implications. The American Surgeon. 2023;**89**(1):25–30. DOI: 10.1177/00031348221101488

[23] Huang CH, Chen PR, Lue KH, Hsieh TC, Chou YF. Evaluation of sarcopenia, frailty, and inflammation on adverse events and survival outcomes in patients with oral cavity squamous cell carcinoma under adjuvant chemoradiotherapy. Journal of Personalized Medicine. 2021;**11**(9):936. DOI: 10.3390/jpm11090936

[24] Molenaar CJL, Papen-Botterhuis NE, Herrle F, Slooter GD. Prehabilitation, making patients fit for surgery - a new frontier in perioperative care. Innovative Surgical Sciences. 2019;**4**(4):132–138. DOI: 10.1515/iss-2019-0017

[25] Nardone V, Marmorino F, Germani MM, Cichowska-Cwalińska N, Menditti VS, Gallo P, Studiale V, Taravella A, Landi M, Reginelli A, Cappabianca S, Girnyi S, Cwalinski T, Boccardi V, Goyal A, Skokowski J, Oviedo RJ, Abou-Mrad A, Marano L. The role of artificial intelligence on tumor boards: Perspectives from surgeons, medical oncologists and radiation oncologists. Current Oncology. 2024;**31**(9):4984–5007. DOI: 10.3390/curroncol31090369

[26] Celi LA, Cellini J, Charpignon ML, Dee EC, Dernoncourt F, Eber R, Mitchell WG, Moukheiber L, Schirmer J, Situ J, Paguio J, Park J, Wawira JG, Yao S. for MIT Critical Data. Sources of bias in artificial intelligence that perpetuate healthcare disparities-A global review. PLOS Digital Health. 2022;**1**(3):e0000022. DOI: 10.1371/journal.pdig.0000022

[27] de Moraes FM, Adissy ENB, Rocha E, et al. Multimodal monitoring intracranial pressure by invasive and noninvasive means. Sci Rep. 2023;**13**(1):18404. DOI: 10.1038/s41598-023-45834-5

[28] Zhang CY, Yang YS, Pei MQ, Chen XL, Chen WC, He HF. The association of cerebral oxygen desaturation with postoperative cognitive dysfunction in older patients: A review. Clin Interv Aging. 2024;**19**:1067–1078. DOI: 10.2147/CIA.S462471

[29] Watson JB, Quintero-Peña C, Moise AC, Lumsden AB, Corr SJ. From data to decision: A comprehensive review of real-time analytics and smart technologies in the surgical suite. Methodist DeBakey Cardiovascular Journal. 2025;**21**(5):5–15. DOI: 10.14797/mdcvj.1658

[30] Putica A, Khanna R, Bosl W, Saraf S, Edgcomb J. Ethical decision-making for AI in mental health: The Integrated Ethical Approach for Computational Psychiatry (IEACP) framework. Psychological Medicine. 2025;**55**:e213. DOI: 10.1017/S0033291725101311

[31] Park D, So K, Prabhakar SK, Kim C, Lee JJ, Sohn JH, Kim JH, Lee SH, Won DO. Early warning score and feasible complementary approach using artificial intelligence-based bio-signal

monitoring system: A review. Biomedical Engineering Letters. 2025;**15**(4):717–734. DOI: 10.1007/s13534-025-00486-4

[32] Zhu L, Wei S, An Y, Hu W, Xie X. Mechanism, contributing factors, and coping strategies of alarm fatigue in intensive care nursing: A qualitative study. Frontiers in Public Health. 2025;**13**:1654389. DOI: 10.3389/fpubh.2025.1654389

[33] Ha EJ, Oh J. Neurocritical care and neurotrauma. Korean Journal of Neurotrauma. 2023;**19**(2):126–127. DOI: 10.13004/kjnt.2023.19.e23

[34] Park DJ, Baik SM, Hong KS, Yi H, Lee JG, Lee JM. Development and external validation of an artificial intelligence model for predicting mortality and prolonged ICU stay in postoperative critically ill patients: A retrospective study. World Journal of Emergency Surgery. 2025;**20**(1):79. DOI: 10.1186/s13017-025-00650-2. Erratum in: World J Emerg Surg. 2025 Nov 26;**20**(1):88. DOI: 10.1186/s13017-025-00663-x

[35] Feinkohl I. Post-Operative cognitive impairment: A cognitive epidemiology perspective. Journal of Intelligence. 2022;**10**(1):18. DOI: 10.3390/jintelligence10010018

[36] Mahajan A, Esper S, Oo TH, et al. Development and validation of a machine learning model to identify patients before surgery at high risk for postoperative adverse events. JAMA Network Open. 2023;**6**(7):e2322285. DOI: 10.1001/jamanetworkopen.2023.22285

[37] Turfa J, Hijazi A, Fadlallah Y, El-Harati M, Dimassi H, Najjar ME. Predictors of 30-Day Mortality and Morbidity Following Craniotomy for Traumatic Brain Injury: An ACS NSQIP Database Analysis. Neurotrauma Reports. 2024;**5**(1):660–670. DOI: 10.1089/neur.2024.0039

[38] Marek K, Zielińska-Nowak E, Redlicka J, Starosta M, Miller E. Intracranial hemorrhage-is very early rehabilitation safe? A narrative review. Journal of Clinical Medicine. 2024;**13**(13):3776. DOI: 10.3390/jcm13133776

[39] Prieto L, Sacristán JA. Problems and solutions in calculating quality-adjusted life years (QALYs). Health and Quality of Life Outcomes. 2003;**1**:80. DOI: 10.1186/1477-7525-1-80

[40] Hazra A, Gogtay N. Biostatistics Series Module 9: Survival analysis. Indian Journal of Dermatology. 2017;**62**(3):251–257. DOI: 10.4103/ijd.IJD_201_17

[41] Maas AIR, Menon DK, Manley GT, et al. Traumatic brain injury: Progress and challenges in prevention, clinical care, and research. The Lancet. Neurology. 2022;**21**(11):1004–1060. DOI: 10.1016/S1474-4422(22)00309-X. Epub 2022 Sep 29. Erratum in: Lancet Neurol. 2022 Dec;**21**(12)e10. DOI: 10.1016/S1474-4422(22)00411-2

[42] Ahmed MM, Okesanya OJ, Olaleke NO, Adigun OA, Adebayo UO, Oso TA, Eshun G, Lucero-Prisno DE3. Integrating digital health innovations to achieve universal health coverage. Promoting Health Outcomes and Quality through Global Public Health Equity. Healthcare (Basel). 2025;**13**(9):1060. DOI: 10.3390/healthcare13091060

[43] Chakradeo K, Huynh I, Balaganeshan SB, Dollerup OL, Gade-Jørgensen H, Laupstad SK, Malham M, Nguyen TL, Hulman A, Varga TV. Navigating fairness aspects of clinical prediction models. BMC Medicine. 2025;**23**(1):567. DOI: 10.1186/s12916-025-04340-3

[44] Cho H, Froelicher D, Dokmai N, Nandi A, Sadhuka S, Hong MM, Berger B. Privacy-Enhancing technologies in biomedical data science. Annu Rev Biomed Data Sci. 2024;**7**(1):317–343. DOI: 10.1146/annurev-biodatasci-120423-120107

[45] Fahim YA, Hasani IW, Kabba S, Ragab WM. Artificial intelligence in healthcare and medicine: Clinical applications, therapeutic advances, and future perspectives. Eur J Med Res. 2025;**30**(1):848. DOI: 10.1186/s40001-025-03196-w

[46] Alfiras MII, Emran AQ, Mohamed AM. Ethics and governance of generative AI in education: A systematic review on responsible adoption. Discov Educ. 2026;**5**:37. DOI: 10.1007/s44217-025-01051-y

[47] Krahe MA, Baker S, Woods L, Larkins SL. Factors that influence digital health implementation in rural, regional, and remote Australia: An overview of reviews and recommended strategies. Aust J Rural Health. 2025;**33**(2):e70045. DOI: 10.1111/ajr.70045

[48] Hu S, Song D, Wan S, Zhang S, Luo C, Li N, Liu G, da Graça Espírito Santo Vasconcelos J, de Carvalho LLC, Neobísi E, da Costa MLB, Etchu Takounjou J, Neves KMD, Dos Ramos da Conceição L, da Costa Encarnação M, Zhao LY. Digital health: Current applications, challenges, and future directions for enhancing healthcare quality and safety. Front Public Health. 2025;**13**:1646802. DOI: 10.3389/fpubh.2025.1646802

[49] Iqbal J, Naseem A, Bashir MA, Yangi K, Bozkurt I, Chaurasia B. Global perspective of neurosurgery practice in lower middle-income countries: Challenges, opportunities, and the path forward. Ann Med Surg (Lond). 2025;**87**(5):2532–2536. DOI: 10.1097/MS9.0000000000003221

[50] Kiosia A, Boylan S, Retford M, Marques LP, Bueno FTC, Kirima C, Islam MS, Naheed A, Wozencraft A. Current data science capacity building initiatives for health researchers in LMICs: Global & regional efforts. Front Public Health. 2024;**12**:1418382. DOI: 10.3389/fpubh.2024.1418382

[51] Chisolm DJ, Dugan JA, Figueroa JF, Lane-Fall MB, Roby DH, Rodriguez HP, Ortega AN. Improving health equity through health care systems research. Health Serv Res. 2023;**58**(Suppl 3): 289–299. DOI: 10.1111/1475-6773.14192

[52] Marwaha JS, Landman AB, Brat GA, Dunn T, Gordon WJ. Deploying digital health tools within large, complex health systems: Key considerations for adoption and implementation. NPJ Digit Med. 2022;**5**(1):13. DOI: 10.1038/s41746-022-00557-1

[53] Barbosa da Silva JJ, Birminghamm M, Rivière Cinnamond A, Boisson E, Valdez ML, Garcia SS, Souza JP, Marti MC, Richards LM, Guzman J, Nelson J, Pesce K, Haddad AE, Fitzgerald J, Bascolo E, D'Agostino M. Bias-Mitigated AI as a foundation for resilient and effective health systems. JMIR Public Health Surveill. 2026;**12**: e88457. DOI: 10.2196/88457

[54] Hirani R, Podder D, Stala O, Mohebpour R, Tiwari RK, Etienne M. Strategies to reduce hospital length of stay: Evidence and challenges. Medicina (Kaunas). 2025;**61**(5):922. DOI: 10.3390/medicina61050922

[55] Alderman JE, Palmer J, Laws E, et al. Tackling algorithmic bias and promoting transparency in health datasets: The STANDING Together consensus recommendations. Lancet Digit Health. 2025;7(1):e64–e88. DOI: 10.1016/S2589-7500(24)00224-3